**TECHNICAL JAPANESE SUPPLEMENTS**
*James L. Davis, General Editor*



# Biotechnology

*James L. Davis*
Department of Engineering Professional Development
University of Wisconsin-Madison

**The University of Wisconsin Press**
**University of Tokyo Press**

The University of Wisconsin Press
114 North Murray Street
Madison, Wisconsin 53715
USA

3 Henrietta Street
London WC2E 8LU
England

ISBN 0-299-14714-2

University of Tokyo Press
7-3-1 Hongo, Bunkyo-ku
Tokyo 113
Japan

ISBN 4-13-087053-X



5 4 3 2 1

Printed in the United States of America

# Preface

This book has been designed as a companion to the textbook *Basic Technical Japanese*, by E. E. Daub, R. B. Bird, and N. Inoue, University of Wisconsin Press and University of Tokyo Press (1990), to which we will refer throughout this volume as "BTJ." After using BTJ as a text for teaching technical Japanese to engineers and scientists we quickly recognized that professionals in different fields would benefit from structured exposure to additional KANJI and vocabulary related to their various specialties. Rather than create several different versions of BTJ, we decided to use BTJ as a core text, and produce a series of supplements. This supplement expands upon the material presented in BTJ, focusing on biotechnology, and is designed to be used in conjunction with BTJ. Biotechnology is itself a broad field. In this instance we consider biotechnology to involve the use or modification of organisms or their constituent materials to solve problems. In this sense biotechnology extends into not only the realms of zoology, botany, bacteriology, microbiology, genetics, biochemistry, biophysics and molecular biology, but also involves applications in agriculture, medicine and the production of pharmaceuticals and other chemicals.

In this volume we present 100 KANJI, not introduced in BTJ, that are used frequently or represent important concepts in some facet of biotechnology in Japanese. These KANJI have been selected based upon the frequency with which they appear in twelve key Japanese references related to biotechnology. Those frequencies were then weighted (subjectively) according to the importance of the terms in which the KANJI appeared. Using these 100 KANJI and the 365 KANJI introduced in BTJ we have compiled extensive lists of Japanese terms that appear in documents related to biotechnology. Exercises and readings, both short and moderate in length, allow the reader to learn the vocabulary of biotechnology in Japanese in as realistic a setting as possible. We assume that before using this book the reader has completed Chapters 1–10 of BTJ and has learned both the KANJI and grammar presented in those chapters. In addition, because the KANJI presented in Chapter 20 of BTJ are particularly important in biological work, these KANJI are also used throughout this book without special note. We assume that the reader has already studied these KANJI and is familiar with many of the terms that contain them. If the reader has not yet learned the KANJI in Chapter 20, the reader should do so before proceeding any further.

We begin with Lesson 0, which provides a review of KATAKANA, HIRAGANA and the KANJI introduced in Chapters 1–10 and 20 of BTJ. With this foundation the reader is ready to explore new KANJI and acquire a broader vocabulary. In each of the 10 subsequent lessons we introduce 10 new KANJI. On the first page of each lesson we provide the major ON and KUN readings and meanings for each KANJI. A list of important terms containing each of the new KANJI follows. For each KANJI the two or three terms that are particularly important or are especially likely to be encountered in technical documents have been indicated with a star. Any term that contains two or more of the 100 KANJI introduced in this book appears in the list associated with each of those KANJI. Some terms in these lists contain KANJI that have not been formally introduced up to that point. All of these KANJI are underlined. Many (but not all) of the underlined KANJI will be introduced at some point in the book. Any KANJI that is not underlined has been formally presented either in BTJ or in this book, and the reader with questions about any of these KANJI should refer to the pertinent chapter for review. A supplementary list of terms that can be formed with the KANJI from the corresponding chapter in BTJ is also included. The inclusion of these lists allows us to present many words that are commonly used in biotechnology and contain the BTJ KANJI but not the 100 introduced in this book. All of these lists should be studied carefully, since these words will show the reader how each of the KANJI is used. Words that appear in these lists will be used later in the book without further explanation. Following the vocabulary lists the reader will find a variety of exercises that will aid the reader with vocabulary recognition, KANJI recognition, grammar review and translation skills. Words that the reader has not yet encountered are listed following the exercise in which the word first appears. The reader may wish to compile a personalized glossary of terms that seem to be particularly useful in the reader's own specialty. The reader should work through the exercises in each lesson carefully before proceeding to the next lesson. Reference to BTJ will provide answers to questions related to grammar. Following Lesson 10 we have included Appendix A, a list of 35 supplementary KANJI, their readings, and some example words. The combination of the 365 KANJI in BTJ, the 100 principal KANJI in this book and these additional 35 KANJI provides the reader with a grand total of 500 KANJI, an excellent foundation for reading technical Japanese. Appendix B contains indexes of ON readings and KUN readings for the 135 KANJI formally presented in this volume.

The lessons of this book have been integrated into the flow of material in BTJ as follows:

| Chapters in BTJ: | 1–10, 20 | 11 | 12 | 13 | 14 | 15 | 16 | 17 | 18 | 19 | — |
|---|---|---|---|---|---|---|---|---|---|---|---|
| Lessons in this book: | 0 | 1 | 2 | 3 | 4 | 5 | 6 | 7 | 8 | 9 | 10 |
| Page numbers in this book: | 1 | 7 | 23 | 37 | 51 | 63 | 79 | 91 | 107 | 119 | 135 |

Thus, when working through Lesson 1 the reader is expected to know all the KANJI and grammar given in the first 11 chapters of BTJ, as well as the KANJI of Chapter 20 in BTJ. Much of the material presented in this volume has been taken, often with modification, from these sources:

廣川書店; ドーランド図説医学大辞典; 廣川書店; 1980
山田常雄, 前川文夫, 江上不二夫, 八杉竜一, 小関治男, 古谷雅樹, 日高敏隆; 生物学辞典; 岩波書店; 1983
福井三郎, 田中渥夫; バイオリアクター; 共立出版; 1986
日本造園学会; 学術用語集農学編; 日本学術振興会; 1986
日本生化学会; 生化学用語辞典; 東京化学同人; 1987
太田一男; バイオテクノロジーの用語; 聖文社; 1987
Inter Press Corp.; バイオテクノロジー用語; Inter Press Corp.; 1987
日本動物学会; 学術用語集動物学編; 丸善; 1988
冨田房男; 産業用バイオテクノロジー辞典; 講談社サイエンティフィク; 1989
日本植物学会; 学術用語集植物学編; 丸善; 1990
今堀和友, 山川民夫; 生化学辞典; 東京化学同人; 1990
橋爪裕司; 分子遺伝学の方法; 学会出版センター; 1991

Dr. Ken Lunde of Adobe Systems, Inc., Mr. Mitsuo Fujita of Daicel Chemical Industries, Ltd., Dr. Robert Fujimura of Oak Ridge National Laboratory (Ret.), and Associate Professor Shingo Kawai of Gifu University examined the manuscript and suggested many useful changes. Professors R. Byron Bird and Edward E. Daub have provided advice and encouragement on this and other projects related to technical Japanese for more than a decade. Salary support from the United States Air Force Office of Scientific Research through the U.S.-Japan Industry and Technology Management Training Program at the University of Wisconsin-Madison is gratefully acknowledged. This volume is dedicated to my wife, Zhen, and to our daughter, Ruth, without whose continuing support it would never have been completed.

James L. Davis
Madison, Wisconsin
October 1994

# LESSON 0　　第0課　　BTJ 1-10 & 20

## KANA AND KANJI REVIEW

This lesson will give you the opportunity to review KATAKANA (Chapter 3), HIRAGANA (Chapter 4) and the KANJI (Chapters 5–10 and 20) that you have learned in *Basic Technical Japanese*. All of these terms are important in some aspect of biotechnology. KATAKANA are particularly important, since the names of most enzymes are written in KATAKANA.

### Ex. 0.1 Enzyme names written in KATAKANA

Pronounce each Japanese term without looking at the English equivalent, then identify the English name of the enzyme.

| | |
|---|---|
| アスパルターゼ | aspartase |
| アミダーゼ | amidase |
| アミラーゼ | amylase |
| インベルターゼ | invertase |
| ウリカーゼ | uricase |
| ウレアーゼ | urease |
| エキソヌクレアーゼ | exonuclease |
| エキソペプチダーゼ | exopeptidase |
| エステラーゼ | esterase |
| エノラーゼ | enolase |
| エピメラーゼ | epimerase |
| エンドグリコシダーゼ | endoglycosidase |
| エンドヌクレアーゼ | endonuclease |
| オキシゲナーゼ | oxygenase |
| カタラーゼ | catalase |
| ガラクトシダーゼ | galactosidase |
| カルボキシラーゼ | carboxylase |
| キシラナーゼ | xylanase |
| キチナーゼ | chitinase |
| キナーゼ | kinase |
| グリコシダーゼ | glycosidase |
| グルカナーゼ | glucanase |
| グルコシダーゼ | glucosidase |
| クロロフィラーゼ | chlorophyllase |
| サッカラーゼ | saccharase |
| ジオキシゲナーゼ | dioxygenase |
| ジスムターゼ | dismutase |
| ジホスファターゼ | diphosphatase |
| スクラーゼ | sucrase |
| セルラーゼ | cellulase |
| デキストラナーゼ | dextranase |
| トリプトファナーゼ | tryptophanase |
| ニトロゲナーゼ | nitrogenase |
| ヌクレアーゼ | nuclease |
| ヌクレオシダーゼ | nucleosidase |
| ヌクレオチダーゼ | nucleotidase |
| ヒドロゲナーゼ | hydrogenase |
| ヒドロペルオキシダーゼ | hydroperoxidase |
| フェノラーゼ | phenolase |
| プロテアーゼ | protease |
| プロテイナーゼ | proteinase |
| ペクチナーゼ | pectinase |
| ペニシリナーゼ | penicillinase |
| ペプチダーゼ | peptidase |
| ペルオキシダーゼ | peroxidase |
| ホスファターゼ | phosphatase |
| ホスホリパーゼ | phospholipase |
| ホスホリラーゼ | phosphorylase |
| ポリメラーゼ | polymerase |
| ホルミラーゼ | formylase |
| マルターゼ | maltase |
| メチラーゼ | methylase |
| ラクターゼ | lactase |
| リアーゼ | lyase |
| リガーゼ | ligase |
| リパーゼ | lipase |

**Ex. 0.2 Important terms written in KATAKANA**

Pronounce each Japanese term without looking at the English equivalent, then identify the English term.

| | |
|---|---|
| アガロース | agarose |
| アデノシル | adenosyl |
| アミロース | amylose |
| アミロペクチン | amylopectin |
| アルギニン | arginine |
| アンチコドン | anticodon |
| イニシエーター | initiator |
| インスリン | insulin |
| インフルエンザウイルス | influenza virus |
| ウイルス | virus |
| オペレーター | operator |
| オペロン | operon |
| オルガネラ | organelle |
| クチクラ | cuticle |
| グリコーゲン | glycogen |
| グリコシド | glycoside |
| グルカン | glucan |
| グルコース | glucose |
| クローニング | cloning |
| グロブリン | globulin |
| ゲノム | genome |
| ゲル | gel |
| コドン | codon |
| コロニー | colony |
| サプレッサー | suppressor |
| シグナル | signal |
| ジフテリア | diptheria |
| シャーレ | Petri dish |
| シリカゲル | silica gel |
| スプライシング | splicing |
| ゼラチン | gelatin |
| デンプン | starch |
| テンペレートファージ | temperate phage |
| ヌクレオチド | nucleotide |
| バイオアッセイ | bioassay |
| バイオセンサー | biosensor |
| バイオリアクター | bioreactor |
| ハイブリッド | hybrid |
| バクテリオファージ | bacteriophage |
| ピリミジン | pyrimidine |
| ビルレントファージ | virulent phage |
| ファージ | phage |
| フィブリノーゲン | fibrinogen |
| フィブリン | fibrin |
| フェリチン | ferritin |
| ブイヨン | bouillon |
| プラーク | plaque |
| プラスミド | plasmid |
| プロセシング | processing |
| プロトプラスト | protoplast |
| プロファージ | prophage |
| プロモーター | promoter |
| フレームシフト | frameshift |
| ベクター | vector |
| ヘパリン | heparin |
| ペプチド | peptide |
| ペルオキシソーム | peroxisome |
| ポテンシャル | potential |
| ホモロジー | homology |
| ポリアクリルアミド | polyacrylamide |
| ポリオウイルス | polio virus |
| ホリデイモデル | Holliday model |
| ホルモン | hormone |
| マクロファージ | macrophage |
| ミトコンドリア | mitochondria |
| メチオニン | methionine |
| リガンド | ligand |
| リソソーム | lysosome |
| リプレッサー | repressor |
| リボース | ribose |
| リボソーム | ribosome |
| レオウイルス | reovirus |
| レトロウイルス | retrovirus |
| レプリコン | replicon |
| ロット | lot, batch |
| ワクチン | vaccine |

### Ex. 0.3 Words frequently written in HIRAGANA

A few technical terms may be written entirely in HIRAGANA, especially if one of the KANJI formerly used for such a term is not a JOOYOO KANJI. Pronounce each Japanese term, and notice the KANJI that may be used as an alternative. Some of these KANJI will be introduced in this book. Some writers may express these terms in KATAKANA, rather than HIRAGANA.

| | | |
|---|---|---|
| かくはん | 【撹拌】 | stirring |
| がん | 【癌】 | cancer |
| けた | 【桁】 | lamella |
| けん | 【腱】 | tendon |
| てんかん | 【癲癇】 | epilepsy |
| ねじりばかり | 【捻り秤】 | torsion balance |
| ばらつき | | dispersion, scatter [in data] |
| みつ | 【蜜】 | nectar |
| らせん | 【螺旋】 | helix |
| ろう | 【蝋】 | wax |

### Ex. 0.4 Words written with KATAKANA and KANJI

Practice reading the following words that contain both KATAKANA and KANJI. You should be able to recognize all of the KANJI that appear in this list.

| | | |
|---|---|---|
| アスパラギン酸 | アスパラギンサン | aspartic acid |
| アポ酵素 | アポコウソ | apoenzyme |
| アミノ酸 | アミノサン | amino acid |
| アルキル化 | アルキルカ | alkylation |
| アロステリック部位 | アロステリックブイ | allosteric site |
| イオン対 | イオンツイ | ion pair |
| イソ酵素 | イソコウソ | isozyme |
| 一次リソソーム | イチジリソソーム | primary lysosome |
| カルボン酸 | カルボンサン | carboxylic acid |
| ガンマ線 | ガンマセン | gamma ray |
| キメラ動物 | キメラドウブツ | chimera animal |
| ギャップ結合 | ギャップケツゴウ | gap junction |
| クエン酸 | クエンサン | citric acid |
| グリコーゲン分解 | グリコーゲンブンカイ | glycogenolysis |
| グルタミン酸 | グルタミンサン | glutamic acid |
| クローン化 | クローンカ | cloning |
| 酵素センサー | コウソセンサー | enzyme sensor |
| ゴルジ体 | ゴルジタイ | Golgi body |
| 成長ホルモン | セイチョウホルモン | growth hormone |
| 性フェロモン | セイフェロモン | sex pheromone |
| タンパク質 | タンパクシツ | protein |
| 電場ジャンプ法 | デンばジャンプホウ | electric field jump method |

| | | |
|---|---|---|
| 電場パルス法 | デンばパルスホウ | electric field pulse method |
| 同質ゲノム | ドウシツゲノム | isogenome |
| ピルビン酸 | ピルビンサン | pyruvic acid |
| ペプチド結合 | ペプチドケツゴウ | peptide bond |
| ホルミル化 | ホルミルカ | formylation |
| メタン発酵 | メタンハッコウ | methane fermentation |
| 溶原性ファージ | ヨウゲンセイファージ | lysogenic phage, temperate phage |
| リソゾーム酵素 | リソゾームコウソ | lysosomal enzyme |
| リン酸 | リンサン | phosphoric acid |
| リン酸化 | リンサンカ | phosphorylation |

**Ex. 0.5 Words written with HIRAGANA and KANJI**

Practice reading aloud the following words. The number of words in this category is not as large as the number of words written with KATAKANA and KANJI, but these words are also important.

| | | |
|---|---|---|
| うつ状態 | ウツジョウタイ | depression |
| かぎ酵素 | かぎコウソ | key enzyme |
| きょう膜 | キョウマク | capsule |
| けん化 | ケンカ | saponification |
| たんぱく質 | タンパクシツ | protein |
| 発がん性 | ハツガンセイ | carcinogenic |
| ろ過 (or ロ過) | ロカ | filtration |

**Ex. 0.6 Words written with the KANJI of Chapters 5-10 and 20 of BTJ**

Many of the following words are essential if you are to read Japanese books or documents pertaining to biotechnology. The number of terms that you can already recognize is quite large indeed. If you do not rember these KANJI, you should review the readings and meanings that are given in *Basic Technical Japanese*. The numbers in the last column indicate the chapters in BTJ where the KANJI used in a particular word were introduced. The terms in this list will reappear without explanation throughout this book, so you should make a point of memorizing them now.

| | | | |
|---|---|---|---|
| 解重合 | カイジュウゴウ | depolymerization | 9,7,8 |
| 加水分解酵素 | カスイブンカイコウソ | hydrolase | 10,5,6,9,20,7 |
| 血液 | ケツエキ | blood | 20,6 |
| 血流 | ケツリュウ | blood flow | 20,7 |
| 原位置 | ゲンイチ | *in situ* | 6,9,10 |
| 固定 | コテイ | fixation | 6,6 |
| 固定化 | コテイカ | immobilization | 6,6,6 |
| 固定化酵素 | コテイカコウソ | immobilized enzyme | 6,6,6,20,7 |
| 固定相 | コテイソウ | stationary phase | 6,6,10 |
| 光合成 | コウゴウセイ | photosynthesis | 10,8,8 |

| | | | |
|---|---|---|---|
| 合成酵素 | ゴウセイコウソ | synthase, synthetase | 8,8,20,7 |
| 酵素 | コウソ | enzyme | 20,7 |
| 酵素学 | コウソガク | enzymology | 20,7,6 |
| 細菌 | サイキン | bacterium | 9,20 |
| 再生 | サイセイ | regeneration | 20,5 |
| 酸化酵素 | サンカコウソ | oxidase | 7,6,20,7 |
| 重合酵素 | ジュウゴウコウソ | polymerase | 7,8,20,7 |
| 触媒 | ショクバイ | catalyst | 20,20 |
| 神経 | シンケイ | nerve | 20,20 |
| 水解小体 | スイカイショウタイ | lysosome | 5,9,5,6 |
| 水酸化酵素 | スイサンカコウソ | hydroxylase | 5,7,6,20,7 |
| 生化学 | セイカガク | biochemistry | 5,6,6 |
| 生合成 | セイゴウセイ | biosynthesis | 5,8,8 |
| 精子 | セイシ | spermatozoon, sperm | 20,5 |
| 生殖 | セイショク | reproduction | 5,20 |
| 生体高分子 | セイタイコウブンシ | biopolymer | 5,6,5,6,5 |
| 生体触媒 | セイタイショクバイ | biocatalyst | 5,6,20,20 |
| 生体物質 | セイタイブッシツ | biological substance | 5,6,6,8 |
| 生体膜 | セイタイマク | biomembrane | 5,6,20 |
| 生物学 | セイブツガク | biology | 5,6,6 |
| 生物体 | セイブツタイ | organism | 5,6,6 |
| 生物発光 | セイブツハッコウ | bioluminescence | 5,6,8,10 |
| 生物物理学 | セイブツブツリガク | biophysics | 5,6,6,9,6 |
| 染色 | センショク | stain(ing) | 20,9 |
| 染色質 | センショクシツ | chromatin | 20,9,8 |
| 染色体 | センショクタイ | chromosome | 20,9,6 |
| 染色分体 | センショクブンタイ | chromatid | 20,9,6,6 |
| 増殖 | ゾウショク | growth, multiplication, propagation | 10,20 |
| 相同 | ソウドウ | homology | 10,7 |
| 相同染色体 | ソウドウセンショクタイ | homologous chromosome | 10,7,20,9,6 |
| 体液性 | タイエキセイ | humoral | 6,6,8 |
| 対合 | タイゴウ | pairing, synapsis | 8,8 |
| 単位膜 | タンイマク | unit membrane | 9,9,20 |
| 炭水化物 | タンスイカブツ | carbohydrate | 7,5,6,6 |
| 単相体 | タンソウタイ | haploid | 9,10,6 |
| 単分子膜 | タンブンシマク | monolayer | 9,6,5,20 |
| 単量体 | タンリョウタイ | monomer | 9,8,6 |
| 単量体酵素 | タンリョウタイコウソ | monomeric enzyme | 9,8,6,20,7 |
| 中間体 | チュウカンタイ | intermediate | 6,6,6 |
| 中心体 | チュウシンタイ | centrosome | 6,9,6 |
| 注入 | チュウニュウ | injection | 7,5 |
| 定数 | テイスウ | constant | 6,8 |

| | | | |
|---|---|---|---|
| 定量法 | テイリョウホウ | assay | 6,8,8 |
| 等位染色体 | トウイセンショクタイ | isochromosome | 7,9,20,9,6 |
| 同位体 | ドウイタイ | isotope | 7,9,6 |
| 動原体 | ドウゲンタイ | centromere | 6,6,6 |
| 同定 | ドウテイ | identification | 7,6 |
| 等電点 | トウデンテン | isoelectric point | 7,8,10 |
| 二量体 | ニリョウタイ | dimer | 5,8,6 |
| 二量体酵素 | ニリョウタイコウソ | dimeric enzyme | 5,8,6,20,7 |
| 発光 | ハッコウ | luminescence | 8,10 |
| 発酵 | ハッコウ | fermentation | 8,20 |
| 発生 | ハッセイ | development | 8,5 |
| 発生学 | ハッセイガク | embryology | 8,5,6 |
| 発生生物学 | ハッセイセイブツガク | developmental biology | 8,5,5,6,6 |
| 比色計 | ヒショクケイ | colorimeter | 6,9,9 |
| 表面 | ヒョウメン | surface | 7,5 |
| 分化 | ブンカ | differentiation | 6,6 |
| 分解 | ブンカイ | degradation, disintegration, decomposition | 6,9 |
| 分光測定 | ブンコウソクテイ | spectrometry | 6,10,10,6 |
| 分子 | ブンシ | molecule | 6,5 |
| 分子間力 | ブンシカンリョク | intermolecular force | 6,5,6,5 |
| 分子式 | ブンシシキ | molecular formula | 6,5,8 |
| 分子生物学 | ブンシセイブツガク | molecular biology | 6,5,5,6,6 |
| 分子量 | ブンシリョウ | molecular weight | 6,5,8 |
| 分染[法] | ブンセン[ホウ] | differential staining | 6,20,8 |
| 分裂 | ブンレツ | division, fission | 6,20 |
| 変性 | ヘンセイ | denaturation | 6,8 |
| 無菌 | ムキン | aseptic, sterile | 7,20 |
| 無性生殖 | ムセイセイショク | asexual reproduction | 7,8,5,20 |
| 誘発 | ユウハツ | induction | 20,8 |
| 溶解 | ヨウカイ | lysis, dissolution | 9,9 |
| 溶菌 | ヨウキン | bacteriolysis | 9,20 |
| 溶菌液 | ヨウキンエキ | lysate | 9,20,6 |
| 溶血 | ヨウケツ | hemolysis | 9,20 |
| 溶原化 | ヨウゲンカ | lysogenization | 9,6,6 |
| 溶原[細]菌 | ヨウゲン[サイ]キン | lysogenic bacterium | 9,6,9,20 |
| 溶媒 | ヨウバイ | solvent | 9,20 |
| 卵黄 | ランオウ | yolk | 20,20 |

| | | |
|---|---|---|
| 因 | イン | factor, cause |
| 疫 | エキ | epidemic, plague |
| 遠 | エン | far, remote |
| | とお(い) | far, remote |
| | とお(くの) | distant |
| 還 | カン | returning, circulation |
| 降 | コウ | falling, descending |
| 脱 | ダツ- | de-, removal, elimination |
| 蛋 | タン | egg |
| 沈 | チン | sinking |
| | しず(む) | to sink |
| 補 | ホ | complement, supplement |
| | おぎな(う) | to compensate, supplement |
| 免 | メン | exempt, avoiding |

因 疫

遠 還

降 脱

蛋 沈

補 免

## 因

| | | |
|---|---|---|
| 遺伝要因 | イデンヨウイン | hereditary factor |
| 因子 ★ | インシ | factor |
| 外因性蛋白質 | ガイインセイタンパクシツ | extrinsic protein |
| 外因性要因 | ガイインセイヨウイン | extrinsic factor |
| 環境要因 | カンキョウヨウイン | environmental factor |
| 強度因子 | キョウドインシ | intensive factor |
| 形質転換因子 | ケイシツテンカンインシ | transforming principle |
| 原因 ★ | ゲンイン | cause, factor |
| 細胞接着因子 | サイボウセッチャクインシ | cell adhesion factor |
| 神経成長因子 | シンケイセイチョウインシ | nerve growth factor |
| 性因子 | セイインシ | sex factor |
| 成長因子 | セイチョウインシ | growth factor |
| 増殖因子 | ゾウショクインシ | growth factor |
| 組織因子 | ソシキインシ | tissue factor |
| 転置因子 | テンチインシ | transposable element |
| 内因子 | ナイインシ | intrinsic factor |
| 内因性 | ナイインセイ | endogenous |
| 稔性因子 | ネンセイインシ | fertility factor |
| 分化誘導因子 | ブンカユウドウインシ | differentiation-inducing factor |
| 補因子 | ホインシ | cofactor |
| 補助因子 | ホジョインシ | cofactor |
| 要因 ★ | ヨウイン | factor |
| 律速因子 | リッソクインシ | rate-limiting factor |
| 立体因子 | リッタイインシ | steric factor |

## 疫

| | | |
|---|---|---|
| 遺伝性免疫 | イデンセイメンエキ | inherited immunity |
| 酵素免疫定量法 | コウソメンエキテイリョウホウ | enzyme immunoassay |
| 細胞性免疫 | サイボウセイメンエキ | cellular immunity, cell-mediated immunity |
| 自己免疫 | ジコメンエキ | autoimmunity |
| 自然免疫 | シゼンメンエキ | natural immunity |
| 体液性免疫 | タイエキセイメンエキ | humoral immunity |
| 免疫 ★ | メンエキ | immunity |
| 免疫遺伝学 | メンエキイデンガク | immunogenetics |
| 免疫化 | メンエキカ | immunization |
| 免疫化学 | メンエキカガク | immunochemistry |
| 免疫学 | メンエキガク | immunology |
| 免疫記憶 | メンエキキオク | immunological memory |

| | | |
|---|---|---|
| 免疫機能 | メンエキキノウ | immune function |
| 免疫グロブリン | メンエキグロブリン | immunoglobulin |
| 免疫系 ★ | メンエキケイ | immune system |
| 免疫原 ★ | メンエキゲン | immunogen |
| 免疫原性 | メンエキゲンセイ | immunogenicity |
| 免疫検定法 | メンエキケンテイホウ | immunoassay |
| 免疫抗体 | メンエキコウタイ | immonoantibody |
| 免疫細胞化学 | メンエキサイボウカガク | immunocytochemistry |
| 免疫細胞溶解 | メンエキサイボウヨウカイ | immune cytolysis |
| 免疫性 | メンエキセイ | immunity |
| 免疫生化学 | メンエキセイカガク | immunobiochemistry |
| 免疫生物学 | メンエキセイブツガク | immunobiology |
| 免疫沈降物 | メンエキチンコウブツ | immune precipitate |
| 免疫定量法 | メンエキテイリョウホウ | immunoassay |
| 免疫反応 | メンエキハンノウ | immunoreaction |
| 免疫複合体 | メンエキフクゴウタイ | immune complex |
| 免疫療法 | メンエキリョウホウ | immunotherapy |
| 臨床免疫学 | リンショウメンエキガク | clinical immunology |

## 遠

| | | |
|---|---|---|
| 遠位尿細管 | エンイニョウサイカン | distal tubule |
| 遠心 ★ | エンシン | centrifugation |
| 遠心加速度 | エンシンカソクド | centrifugal acceleration |
| 遠心機 | エンシンキ | centrifuge |
| 遠心性神経 | エンシンセイシンケイ | efferent nerve |
| 遠心[沈殿]する | エンシン[チンデン]する | to centrifuge |
| 遠心[分離]する | エンシン[ブンリ]する | to centrifuge |
| 遠心力場 | エンシンリキば | centrifugal field |
| 遠心力 ★ | エンシンリョク | centrifugal force |
| 遠赤外 ★ | エンセキガイ | far infrared |
| 高速遠心機 | コウソクエンシンキ | high speed centrifuge |
| 蔗糖密度勾配遠心[分離]法 | ショトウミツドコウバイエンシン[ブンリ]ホウ | sucrose density-gradient centrifugation |
| 超遠心機 | チョウエンシンキ | ultracentrifuge |
| 遠縁 | とおエン | distant relation |
| 密度勾配遠心[分離]法 | ミツドコウバイエンシン[ブンリ]ホウ | density-gradient centrifugation |

## 還

| | | |
|---|---|---|
| 還元 ★ | カンゲン | reduction |
| 還元酵素 ★ | カンゲンコウソ | reductase |
| 還元電位 | カンゲンデンイ | reduction potential |
| 還元力 | カンゲンリョク | reducing power |

| | | |
|---|---|---|
| 還流 | カンリュウ | reflux |
| 光還元 | コウカンゲン | photoreduction |
| 酸化還元 ★ | サンカカンゲン | oxidation-reduction |
| 酸化還元酵素 | サンカカンゲンコウソ | oxidoreductase, oxidation-reduction enzyme |
| 酸化還元電位 | サンカカンゲンデンイ | oxidation-reduction potential |
| 触媒還元 | ショクバイカンゲン | catalytic reduction |
| 分子内酸化還元 | ブンシナイサンカカンゲン | internal oxidation-reduction |

## 降

| | | |
|---|---|---|
| 以降 | イコウ | from this point onward |
| 沈降 ★ | チンコウ | precipitation, sedimentation |
| 沈降係数 | チンコウケイスウ | sedimentation coefficient |
| 沈降素 | チンコウソ | precipitin |
| 沈降速度法 | チンコウソクドホウ | sedimentation velocity method |
| 沈降定数 | チンコウテイスウ | sedimentation constant |
| 沈降反応 ★ | チンコウハンノウ | precipitation reaction |
| 沈降物 ★ | チンコウブツ | sediment, precipitate |
| 沈降平衡法 | チンコウヘイコウホウ | sedimentation equilibrium method |
| 定量沈降反応 | テイリョウチンコウハンノウ | quantitative precipitin reaction |
| 氷点降下剤 | ヒョウテンコウカザイ | freezing point depressant |
| 免疫沈降物 | メンエキチンコウブツ | immune precipitate |

## 脱

| | | |
|---|---|---|
| 酸化的脱炭酸反応 | サンカテキダツタンサンハンノウ | oxidative decarboxylation |
| 脱アシル[化] | ダツアシル[カ] | deacylation |
| 脱アミノ[化] | ダツアミノ[カ] | deamination |
| 脱アルキル[化] | ダツアルキル[カ] | dealkylation |
| 脱塩 | ダツエン | desalting, desalination |
| 脱灰 | ダッカイ | decalcification |
| 脱外被 | ダツガイヒ | uncoating |
| 脱核 | ダッカク | enucleation |
| 脱核細胞 | ダッカクサイボウ | enucleated cell |
| 脱気 | ダッキ | degassing |
| 脱共役 | ダツキョウヤク | uncoupling |
| 脱グリコシル[化] | ダツグリコシル[カ] | deglycosylation |
| 脱結合 | ダツケツゴウ | decoupling |

| | | |
|---|---|---|
| 脱色する | ダッショクする | to decolorize |
| 脱水 ★ | ダッスイ | dehydration |
| 脱髄 | ダツズイ | demyelination |
| 脱水酵素 | ダッスイコウソ | dehydratase |
| 脱水素 | ダツスイソ | dehydrogenation |
| 脱水素酵素 | ダツスイソコウソ | dehydrogenase |
| 脱炭酸 ★ | ダツタンサン | decarboxylation |
| 脱炭酸酵素 | ダツタンサンコウソ | decarboxylase |
| 脱着 | ダッチャク | desorption |
| 脱同期 | ダツドウキ | desynchronization |
| 脱分化 | ダツブンカ | dedifferentiation |
| 脱ホルミル酵素 | ダツホルミルコウソ | deformylase |
| 脱メチル[化] | ダツメチル[カ] | demethylation |
| 脱離 ★ | ダツリ | elimination |
| 脱離酵素 | ダツリコウソ | lyase |
| 脱リン酸化 ★ | ダツリンサンカ | dephosphorylation |
| リン酸化-脱リン酸化回路 | リンサンカ-ダツリンサンカカイロ | phosphorylation-dephosphorylation cycle |

# 蛋

| | | |
|---|---|---|
| 塩基性蛋白質 | エンキセイタンパクシツ | basic protein |
| 外因性蛋白質 | ガイインセイタンパクシツ | extrinsic protein |
| 核蛋白質 ★ | カクタンパクシツ | nucleoprotein |
| 結合蛋白質 | ケツゴウタンパクシツ | binding protein |
| 酸性蛋白質 | サンセイタンパクシツ | acidic protein |
| 色素蛋白質 | シキソタンパクシツ | chromoprotein |
| 制御蛋白質 | セイギョタンパクシツ | control protein |
| 相同蛋白質 | ソウドウタンパクシツ | homologous protein |
| 単細胞蛋白質 | タンサイボウタンパクシツ | single cell protein |
| 担体蛋白質 | タンタイタンパクシツ | carrier protein |
| 蛋白質 ★ | タンパクシツ | protein |
| 蛋白質合成 | タンパクシツゴウセイ | protein synthesis |
| 蛋白質生合成 | タンパクシツセイゴウセイ | protein biosynthesis |
| 蛋白質分解 | タンパクシツブンカイ | proteolysis |
| 蛋白質リン酸化酵素 | タンパクシツリンサンカコウソ | protein kinase |
| 糖蛋白質 ★ | トウタンパクシツ | glycoprotein |
| 熱ショック蛋白質 | ネツショックタンパクシツ | heat shock protein |
| 非ヒストン蛋白質 | ヒヒストンタンパクシツ | nonhistone protein |
| 変性蛋白質 | ヘンセイタンパクシツ | denatured protein |
| 膜蛋白質 | マクタンパクシツ | membrane protein |
| 膜内在性蛋白質 | マクナイザイセイタンパクシツ | membrane intrinsic protein |
| リン蛋白質 | リンタンパクシツ | phosphoprotein |

## 沈

| | | |
|---|---|---|
| 遠心[沈殿]する ★ | エンシン[チンデン]する | to centrifuge |
| 沈降 ★ | チンコウ | precipitation, sedimentation |
| 沈降係数 | チンコウケイスウ | sedimentation coefficient |
| 沈降素 | チンコウソ | precipitin |
| 沈降速度法 | チンコウソクドホウ | sedimentation velocity method |
| 沈降定数 | チンコウテイスウ | sedimentation constant |
| 沈降反応 | チンコウハンノウ | precipitation reaction |
| 沈降物 | チンコウブツ | sediment, precipitate |
| 沈降平衡法 | チンコウヘイコウホウ | sedimentation equilibrium method |
| 沈殿 (or 沈澱) ★ | チンデン | precipitation |
| 沈殿[物] | チンデン[ブツ] | precipitate |
| 定量沈降反応 | テイリョウチンコウハンノウ | quantitative precipitin reaction |
| 等電沈殿 | トウデンチンデン | isoelectric precipitation |
| 免疫沈降物 | メンエキチンコウブツ | immune precipitate |

## 補

| | | |
|---|---|---|
| 酵素補充療法 | コウソホジュウリョウホウ | enzyme replacement therapy |
| 相補 ★ | ソウホ | complementation |
| 相補作用 | ソウホサヨウ | complementary action |
| 相補性 | ソウホセイ | complementarity |
| 相補性検定 | ソウホセイケンテイ | complementation test |
| 相補単位 | ソウホタンイ | complon |
| 相補的対合 | ソウホテキタイゴウ | complementary pairing |
| 補因子 ★ | ホインシ | cofactor |
| 補完 | ホカン | complementation |
| 補給 | ホキュウ | replenishment |
| 補欠分子族 | ホケツブンシゾク | prosthetic group |
| 補酵素 ★ | ホコウソ | coenzyme |
| 補助因子 | ホジョインシ | cofactor |
| 補助色素 | ホジョシキソ | accessory pigment |
| 補足 | ホソク | complement, supplement |
| 補足遺伝子 | ホソクイデンシ | complementary gene |
| 補体 | ホタイ | complement |
| 補体結合反応 | ホタイケツゴウハンノウ | complement fixation reaction |

# 免

| | | |
|---|---|---|
| 遺伝性免疫 | イデンセイメンエキ | inherited immunity |
| 酵素免疫定量法 | コウソメンエキテイリョウホウ | enzyme immunoassay |
| 細胞性免疫 | サイボウセイメンエキ | cellular immunity, cell-mediated immunity |
| 自己免疫 | ジコメンエキ | autoimmunity |
| 自然免疫 | シゼンメンエキ | natural immunity |
| 体液性免疫 | タイエキセイメンエキ | humoral immunity |
| 免疫 ★ | メンエキ | immunity |
| 免疫遺伝学 | メンエキイデンガク | immunogenetics |
| 免疫化 | メンエキカ | immunization |
| 免疫化学 | メンエキカガク | immunochemistry |
| 免疫学 | メンエキガク | immunology |
| 免疫記憶 | メンエキキオク | immunological memory |
| 免疫機能 | メンエキキノウ | immune function |
| 免疫グロブリン | メンエキグロブリン | immunoglobulin |
| 免疫系 ★ | メンエキケイ | immune system |
| 免疫原 ★ | メンエキゲン | immunogen |
| 免疫原性 | メンエキゲンセイ | immunogenicity |
| 免疫検定法 | メンエキケンテイホウ | immunoassay |
| 免疫抗体 | メンエキコウタイ | immunoantibody |
| 免疫細胞化学 | メンエキサイボウカガク | immunocytochemistry |
| 免疫細胞溶解 | メンエキサイボウヨウカイ | immune cytolysis |
| 免疫性 | メンエキセイ | immunity |
| 免疫生化学 | メンエキセイカガク | immunobiochemistry |
| 免疫生物学 | メンエキセイブツガク | immunobiology |
| 免疫沈降物 | メンエキチンコウブツ | immune precipitate |
| 免疫定量法 | メンエキテイリョウホウ | immunoassay |
| 免疫反応 | メンエキハンノウ | immunoreaction |
| 免疫複合体 | メンエキフクゴウタイ | immune complex |
| 免疫療法 | メンエキリョウホウ | immunotherapy |
| 臨床免疫学 | リンショウメンエキガク | clinical immunology |

*SUPPLEMENTARY VOCABULARY USING KANJI FROM CHAPTER 11*

外膜　ガイマク　outer membrane
間期　カンキ　interphase, interkinesis
器官　キカン　organ
後期　コウキ　anaphase
酵素前駆体　コウソゼンクタイ　enzyme precursor, proenzyme
前期　ゼンキ　prophase
前駆体　ゼンクタイ　precursor
対数増殖期　タイスウゾウショクキ　logarithmic growth phase
中期　チュウキ　metaphase
内膜　ナイマク　intima
不斉　フセイ　asymmetric
不対合　フタイゴウ　desynapsis
分裂間期　ブンレツカンキ　interphase, interkinesis
分裂期　ブンレツキ　mitotic phase
分裂周期　ブンレツシュウキ　mitotic cycle
無菌状態　ムキンジョウタイ　aseptic condition
卵白　ランパク　albumen

## EXERCISES

### Ex. 1.1 Matching Japanese and English terms

( ) 遠心沈殿
( ) 還元酵素
( ) 酸化還元電位
( ) 成長因子
( ) 相同蛋白質
( ) 脱水素酵素
( ) 脱リン酸化
( ) 蛋白質合成
( ) 沈降定数
( ) 補因子
( ) 補体結合反応
( ) 免疫学
( ) 免疫沈降物

1. cofactor
2. complement fixation reaction
3. dehydrogenase
4. dephosphorylation
5. growth factor
6. homologous protein
7. immune precipitate
8. immunology
9. oxidation-reduction potential
10. protein synthesis
11. reductase
12. sedimentation constant
13. spinning down

### Ex. 1.2 KANJI with similar structural elements

Look carefully at each of the two KANJI on the left, and note which structural element is common to both. Combine each KANJI on the left with the appropriate KANJI on the right to make a meaningful JUKUGO. Each technical term that contains one or more of the 100 KANJI introduced in this book can be found in the vocabulary lists for those KANJI. Other terms can be found in one of the supplementary vocabulary lists, including Lesson 0.

| | (1) | (2) | | |
|---|---|---|---|---|
| 1. | (1) 因 | (2) 固 | 増殖( )子 | ( )定化酵素 |
| 2. | (1) 疫 | (2) 度 | 沈降速( )法 | 免( )定量法 |
| 3. | (1) 遠 | (2) 還 | ( )赤外 | ( )流 |
| 4. | (1) 遠 | (2) 速 | ( )心機 | 律( )因子 |
| 5. | (1) 還 | (2) 置 | 原位( ) | ( )元力 |
| 6. | (1) 脱 | (2) 膜 | 生体( ) | ( )炭酸酵素 |
| 7. | (1) 蛋 | (2) 触 | ( )媒還元 | 変性( )白質 |
| 8. | (1) 沈 | (2) 注 | ( )殿物 | ( )入 |
| 9. | (1) 沈 | (2) 法 | ( )降係数 | 定量( ) |
| 10. | (1) 免 | (2) 色 | ( )疫原 | 染( )体 |

**Ex. 1.3 Matching Japanese technical terms with definitions**

Read each definition carefully, and then choose the appropriate technical term. Words that you have not yet encountered are listed following the definitions.

| | | |
|---|---|---|
| (　) 遠心機 | (　) 蛋白質 | (　) 補酵素 |
| (　) 酸化還元酵素 | (　) 沈降速度法 | (　) 免疫 |
| (　) 成長因子 | (　) 沈降反応 | (　) 免疫グロブリン |
| (　) 脱炭酸酵素 | | |

1. 生物または細胞の数、重量、体積などを増加させる作用をもつ物質。
2. 生体の内部環境に外来性または内因性の異物が存在したとき、それを排除しようとする機構。
3. 抗体およびこれと構造上・機能上の関連性のある蛋白質の総称。
4. 溶液中の溶質の分析や分離の目的で溶液の遠心を行う装置。
5. 酸化還元反応を触媒する酵素。
6. 超遠心機を用いて溶液中の溶質の沈降係数を求める実験法。
7. 抗体と可溶性抗原とが特異的に作用して沈降物を作る反応。
8. カルボン酸のカルボキシル基を炭酸として脱離する酵素の総称。
9. L-α-アミノ酸からなるポリペプチドを主要成分とする高分子物質。
10. アポ酵素と可逆的に結合して酵素作用の発現に寄与する補欠分子族。

| | | | | | |
|---|---|---|---|---|---|
| 細胞 | サイボウ | cell | 分析 | ブンセキ | analysis |
| もつ | 【持つ】 | to have | 分離 | ブンリ | separation |
| 環境 | カンキョウ | environment | 目的 | モクテキ | purpose |
| 外来性 | ガイライセイ | extraneous | 装置 | ソウチ | apparatus |
| 異物 | イブツ | foreign matter | 求める | もとめる | to seek |
| 存在 | ソンザイ | existence | 可溶性 | カヨウセイ | soluble |
| 排除 | ハイジョ | exclusion | 抗原 | コウゲン | antigen |
| しようとする | | to try to do | 特異的 | トクイテキ | specific |
| 機構 | キコウ | mechanism | -基 | キ | radical, group |
| 抗体 | コウタイ | antibody | 主要 | シュヨウ | principal, major |
| 構造 | コウゾウ | structure | 可逆的 | カギャクテキ | reversible |
| 機能 | キノウ | function | 発現 | ハツゲン | expression |
| 関連性 | カンレンセイ | relevance | 寄与 | キヨ | contribution |
| 総称 | ソウショウ | general term | | | |

### Ex. 1.4 Sentence translations

Read each sentence carefully, and then translate it. Words that you have not yet encountered are listed following the sentences.

1. 補因子と活性化物質との間には厳密な定義の区別がないが、活性化物質が人為的な活性化をもたらすのに対して、補因子は本来ある酵素とともにあって活性に必須の因子、つまり補酵素と同義語に近い用語である。
2. 生体の免疫機能を研究する免疫学は、取り扱う対象と方法により免疫生物学、免疫化学、臨床免疫学などの多くの分野に分かれている。
3. 一般に蛋白質の免疫原性が最も強く、生物学的分類で遠縁のものや分子量の大きなものほど免疫原性が強い。抗原の免疫原性は免疫の条件によって変動する。
4. 遠心力による物質の移動を沈降といい、これを用いると核酸や蛋白質のような生体物質、オルガネラ、あるいは細胞自体の調製や分析を行うことができる。
5. 酸化還元反応の様式、性質、供与体や受容体の種類などによって、酸化還元酵素が脱水素酵素、還元酵素、酸化酵素、オキシゲナーゼ、ヒドロペルオキシダーゼなどに分類されている。
6. 沈降定数sの大きさは単位の遠心加速度による移動速度で、時間の次元をもつ。sは溶質の濃度とともに減少し、$s^{-1}$を濃度に対して目盛ると直線を与える。
7. 一定量の抗体に変化量の抗原を加えると、抗原抗体量比のある値で沈降物量が最高となる。これにより反応系に含まれる抗原および抗体のすべてが沈降物となる等量点がでる。
8. 酸化的脱炭酸反応を行う脱炭酸酵素も行わない脱炭酸酵素もある。前者は$NAD(P)^+$依存性の脱水素反応と同時に脱炭酸を伴うものである。
9. 蛋白質分子がほぼ生理的条件下でとっている固有の構造を生とよぶが、種々の原因により一次構造は変化せずに高次構造のみが破壊され、生の状態と物性が変化した蛋白質を変性蛋白質とよぶ。
10. 補体とは新鮮血清中に存在し、免疫複合体に非特異的に反応する因子で、抗原が赤血球や細菌であれば、その溶血や溶菌を起こす物質と定義されてきた。

| | | |
|---|---|---|
| 活性化物質 | カッセイカブッシツ | activator |
| 活性 | カッセイ | activity |
| 厳密な | ゲンミツな | strict |
| 定義 | テイギ | definition |
| 区別 | クベツ | distinction |
| 人為的 | ジンイテキ | artificial |
| もたらす | | to bring about |
| 本来 | ホンライ | naturally |
| -とともに | 【と共に】 | together (with) |
| 必須 | ヒッス | essential |
| 同義語 | ドウギゴ | synonym |
| 用語 | ヨウゴ | term(inology) |
| 研究 | ケンキュウ | research |
| 取り扱う | とりあつかう | to deal with |
| 対象 | タイショウ | subject |
| 分野 | ブンヤ | field (of learning) |
| 一般に | イッパンに | in general |
| 分類 | ブンルイ | classification |
| 条件 | ジョウケン | condition |
| 移動 | イドウ | movement |
| 核酸 | カクサン | nucleic acid |
| 自体 | ジタイ | itself |
| 調製 | チョウセイ | preparation |
| 様式 | ヨウシキ | form |
| 供与体 | キョウヨタイ | donor |
| 受容体 | ジュヨウタイ | acceptor, receptor |
| 種類 | シュルイ | variety, kind |

| | | | | | |
|---|---|---|---|---|---|
| 移動速度 | イドウソクド | migration velocity | 伴う | ともなう | to carry with |
| 濃度 | ノウド | concentration | 生理的 | セイリテキ | physiological |
| 減少 | ゲンショウ | decrease | とる | | to adopt; to take |
| 目盛る | めもる | to mark on a scale, to plot | 種々の | シュジュの | various |
| | | | 一次 | イチジ | primary |
| 与える | あたえる | to give | 高次 | コウジ | higher order |
| 値 | あたい | value | 破壊 | ハカイ | destruction |
| -系 | ケイ | system | 新鮮 | シンセン | fresh, new |
| 等量点 | トウリョウテン | equivalence point | 血清 | ケッセイ | (blood) serum |
| | | | 非特異的 | ヒトクイテキ | nonspecific |
| 前者 | ゼンシャ | the former | 赤血球 | セッケッキュウ | erythrocyte, red blood cell |
| 依存性 | イゾンセイ | dependent | | | |

**Ex. 1.5 Additional dictionary entries**

( ) 酵素
( ) 酵素学
( ) 生化学
( ) 生物物理学
( ) 染色体
( ) 相同染色体
( ) 発酵
( ) 分子生物学
( ) 溶菌
( ) 溶原化

1. 化学の知識を基礎とし、化学的手段によって生命現象を解明する学問。
2. かつては物理学的方法を用いる生理学的研究を指したが、現在では、もっと広い意味で、物理学的な考え方や方法で生命現象を研究する学問。
3. 広義には、有機物質が微生物によって分解される現象を指すが、狭義には、炭水化物が細菌や酵母の酵素によって無酸素的に分解されること。
4. 酵素を対象として生物学的、物理学的、化学的、工学的に研究する学問。
5. 生物の生成する、体内での化学反応をそれぞれ触媒する能力をもつ蛋白質分子。
6. 生命現象を分子、特に生体高分子の構造と機能に基づいて解明しようとする生物学の一分野。
7. テンペレートファージが細菌細胞に感染して、菌染色体中に組込まれるかまたはプラスミドとしてプロファージ状態になること。
8. 動植物細胞が有糸分裂する際に出現し塩基性色素で濃染される構造体。
9. 二倍性の生物において父方及び母方から由来した形態の相等しい一対の染色体。
10. ファージの溶菌感染の際、細胞壁および細胞膜を分解する溶菌酵素が生成され、それにより細胞が溶解し子ファージが放出されること。

| | | | | | |
|---|---|---|---|---|---|
| 知識 | チシキ | knowledge | かつて | | formerly |
| 基礎 | キソ | base | 指す | さす | to indicate |
| 手段 | シュダン | means, measure | 現在 | ゲンザイ | currently |
| 生命 | セイメイ | life | 広い | ひろい | wide |
| 現象 | ゲンショウ | phenomenon | 意味 | イミ | meaning |
| 解明 | カイメイ | elucidation | 広義 | コウギ | broad sense, broad meaning |
| 学問 | ガクモン | field of study | | | |

| | | |
|---|---|---|
| 微生物 | ビセイブツ | microorganism |
| 狭義 | キョウギ | narrow sense, narrow meaning |
| 酵母 | コウボ | yeast |
| 無酸素的 | ムサンソテキ | anaerobic |
| 生成 | セイセイ | production |
| 能力 | ノウリョク | ability, capability |
| 特に | トクに | in particular |
| -に基づいて | にもとづいて | based upon |
| 感染 | カンセン | infection |
| 組込む | くみこむ | to incorporate |
| 動植物- | ドウショクブツ- | plant and animal |
| 有糸分裂 | ユウシブンレツ | mitosis |
| 際(に) | サイ(に) | when |
| 出現 | シュツゲン | appearance |
| 塩基性 | エンキセイ | basic, alkaline |
| 濃染する | ノウセンする | to dye deeply |
| 構造体 | コウゾウタイ | structural body |
| 二倍性の | ニバイセイの | diploid |
| 父方 | ちちかた | father |
| 母方 | ははかた | mother |
| 由来する | ユライする | to be derived from |
| 形態 | ケイタイ | morphology |
| 溶菌感染 | ヨウキンカンセン | lytic infection |
| 細胞壁 | サイボウヘキ | cell wall |
| 細胞膜 | サイボウマク | cytoplasmic membrane |
| 放出する | ホウシュツする | to release |

#### Ex. 1.6 Additional sentence translations

1. ウイルスの体は、主として核酸と蛋白質とからなっている。バクテリオファージでは、その核酸の一種であるDNAが指令を出して、増殖が行われている。ウイルスの種類によっては、別の核酸がDNAと同じような働きをすることも知られている。

2. m-RNAにはA・G・C・Uの四種の塩基があるので、四の三乗で64通りのコドンが存在する。61種のコドンがアミノ酸に対応している。20種のアミノ酸はいくつかのコドンによって指定される。残りの三種のコドンは、アミノ酸に関しては意味がなく、蛋白質合成の終了を示す情報をもっている。

3. DNAは四種類のヌクレオチドが多数結合した鎖状の化合物である。ヌクレオチドはデオキシリボースという糖に、リン酸基と塩基が結合したものである。塩基は弱アルカリ性の有機化合物で、アデニン(A)、シトシン(C)、グアニン(G)、チミン(T)の四種類がある。この塩基の違いによって、ヌクレオチドは四種類になる。

4. 大腸菌の細胞壁についた$T_2$ファージは、頭部にあるDNAを、尾部を注射器のようにして大腸菌内に注入する。注入されたDNAは、大腸菌のDNAを破壊し、自己と同じDNAをつくる。これらのDNAは、$T_2$ファージの体をつくるための蛋白質の種類を決める。ファージ蛋白質が合成され、多数の$T_2$ファージが大腸菌を溶かして、外に出てくる。

5. 細菌類の細胞質の中には、それ自身の持つDNAのほかに、寄生された形で別のDNAが存在している。例えば、大腸菌の性を決定するF因子や薬剤に対する抵抗性に関係のあるR因子などがそれである。他にも多くのこうした因子が発見され、まとめてプラスミドとよばれるようになった。プラスミドの多くは、環状の二本鎖DNAである。

6. 生物はリボソームで蛋白質をつくっている。リボソームは細胞質内に存在する小さな粒子で、二つの部分から構成されている。主要成分は蛋白質とRNAである。大腸菌では、一個体に15,000個以上も含まれ、重量にすると全体の四分の一ぐらいを占めるといわれる。

7. 構造遺伝子の一端には、隣接するようにオペレーター遺伝子、その隣(構造遺伝子とともにオペレーター遺伝子を挟むような位置(にはプロモーター遺伝子が存在する。オペレーター遺伝子はいわば構造遺伝子の機能のスイッチのような働きをする。また、プロモーター遺伝子は、構造遺伝子からm-RNAが合成される時に必要なRNA合成酵素と結合する遺伝子である。

8. 生体内で蛋白質を構成するアミノ酸には、約20種が知られている。基本的には炭素原子に水素・アミノ基・カルボキシル基が共通に結合している。そして、結合するもう一つの原子あるいは原子団によって種類が異なってくる。アスパラギン酸、グルタミン酸、アルギニンなどは、日常生活の中で一度ぐらいは耳にすることのあるアミノ酸である。

9. DNAの一本鎖から構造遺伝子などの塩基配列を転写したm-RNAが合成される反応を促進する酵素は、RNAポリメラーゼである。m-RNAへの転写が起こるときに、まず、RNAポリメラーゼがオペレーター部位の塩基配列を確認するために、特別な蛋白質因子を持っていると考えられている。真核生物では三種のRNAポリメラーゼが知られており、その中の一種がm-RNAの合成に作用しているといわれている。

10. RNAの基本的な構造は、DNAと同様にヌクレオチドが多数結合して、鎖状になっている。構成しているヌクレオチドには、少しの差がある。まず、糖がリボースで、デオキシリボースとは酸素原子がつくかつかないかの違いである。また、四種類の塩基もA・C・G・U(ウラシル)で、チミン(T)の代わりにUが入っているだけである。このように、DNA分子とRNA分子とは、基本的な構造がよく似ている。

| | | |
|---|---|---|
| 主として | シュとして | principally |
| -種 | -シュ | variety |
| 指令 | シレイ | order, instruction |
| 別の | ベツの | (an)other |
| 働き | はたらき | action, working |
| 知る | しる | to know |
| 四 | よん | four |
| 塩基 | エンキ | base |
| 四の三乗 | よんのサンジョウ | four raised to the third power |
| 通り | とおり | kind, sort |
| いくつか | | some number |
| 指定 | シテイ | designation |
| 残り | のこり | remainder |
| 三 | サン | three |
| 終了 | シュウリョウ | termination |
| 情報 | ジョウホウ | information |
| 鎖状 | サジョウ | chain-like |
| 糖 | トウ | sugar |
| リン酸基 | リンサンキ | phosphate group |
| 弱 | ジャク- | weak(ly) |
| 大腸菌 | ダイチョウキン | *Escherichia coli* (*E. coli*) |
| 頭部 | トウブ | head |

| | | |
|---|---|---|
| 尾部 | ビブ | tail |
| 注射器 | チュウシャキ | syringe |
| 自己 | ジコ | self |
| 決める | きめる | to decide |
| 細菌類 | サイキンルイ | bacteria |
| 細胞質 | サイボウシツ | cytoplasm |
| それ自身 | それジシン | itself, oneself |
| 持つ | もつ | to possess |
| 寄生 | キセイ | parasitism |
| 性 | セイ | sex |
| 決定 | ケッテイ | determination |
| 薬剤 | ヤクザイ | drug |
| 抵抗性 | テイコウセイ | resistance |
| まとめて | | collectively |
| よぶ | | to call |
| 環状 | カンジョウ | circular |
| 二本鎖 | ニホンサ | double stranded |
| 粒子 | リュウシ | particle |
| 一個体 | イッコタイ | one (body) |
| 15,000個 | 15,000コ | 15,000 (pieces or units) |
| 四分の一 | よんブンのイチ | one fourth |
| 占める | しめる | to occupy |
| 一端 | イッタン | one end |
| 隣接 | リンセツ | contiguity |
| その隣 | そのとなり | adjacent to that |
| 挟む | はさむ | to put between |
| いわば | | in a manner of speaking |
| 約 | ヤク | approximately |
| 基本的 | キホンテキ | fundamental |
| 共通 | キョウツウ | commonality |
| 原子団 | ゲンシダン | (atomic) group |
| 異なる | ことなる | to differ |
| 日常生活 | ニチジョウセイカツ | daily life |
| 耳にする | みみにする | to hear about |
| 一本鎖 | イッポンサ | single stranded |
| 配列 | ハイレツ | sequence |
| 転写 | テンシャ | transcription |
| 促進 | ソクシン | promotion |
| 部位 | ブイ | site |
| 確認 | カクニン | confirmation |
| 特別な | トクベツな | special |
| 真核生物 | シンカクセイブツ | eucaryote |
| 同様に | ドウヨウに | in the same way |
| 代わり | かわり | substitute |
| 似る | にる | to resemble |

**Translations for Ex. 1.4**

1. There is no distinction between the strict definitions of cofactor and activator. However, an activator brings about artificial activity, while a cofactor naturally exists together with an enzyme and is essential for [the enzyme's] activity. In other words, the term cofactor is nearly synonomous with coenzyme.
2. Immunology, in which we do research about the immune functions of an organism, is divided into many fields, such as immunobiology, immunochemistry and clinical immunology, according to the subject dealt with and the methods [used].
3. In general, the immunogenicity of proteins is the strongest. The more distant the relation in biological classification, or the larger the molecular weight, the stronger is the immunogenicity. The immunogenicity of antibodies varies depending upon the immune conditions.
4. The movement of substances due to centrifugal force is called sedimentation. Using sedimentation it is possible to prepare and analyze biological substances such as nucleic acids or proteins, organelles, or [even] cells themselves.
5. An oxidation-reduction enzyme is classified as a dehydrogenase, a reductase, an oxidase, an oxygenase, a hydroperoxidase, and so forth, according to the form and characteristics of the oxidation-reduction reaction, the variety of the donor and acceptor, and so on.
6. The magnitude of the sedimentation constant $s$ is the migration velocity due to a unit [quantity of] centrifugal acceleration; $s$ carries the dimension of time. $s$ decreases with the concentration of solute; if we plot $s^{-1}$ against concentration, we obtain a straight line.
7. If we add varying quantities of antigen to a fixed quantity of antibody, at a certain antigen-antibody ratio the amount of precipitate will reach a maximum [value]. This is the equivalence point, at which all of the antigen and antibody included in the reaction system becomes part of the precipitate.
8. There are decarboxylases that carry out oxidative decarboxlyation reactions and those that do not. For the former the decarboxylation proceeds simultaneously with the NAD $(P)^+$-dependent dehydrogenation reaction.
9. The characterisitc structure adopted by a protein under nearly physiological conditions is called "native." A protein whose native state and physical properties have changed as a result of the breakdown of the higher order structure due to various factors, even though the primary structure has not changed, is known as a denatured protein.
10. A complement is a factor that exists in fresh (blood) serum and reacts nonspecifically with an immune complex. If the antigen is an erythrocyte or a bacteria, a complement is defined as a substance that causes the antigen's hemolysis or bacteriolysis to occur.

| | | |
|---|---|---|
| 泳 | エイ | swimming, keeping afloat |
| 栄 | エイ | prosperity, thriving |
| 害 | ガイ | injury, harm |
| 照 | ショウ | comparison, illumination |
| 障 | ショウ | obstacle, interference |
| 清 | セイ | clear, pure |
| 阻 | ソ | obstruction, hindrance |
| 培 | バイ | cultivation, nurturing |
| 融 | ユウ | fusing, melting |
| 養 | ヨウ | nourishment, feeding |
| | やしな(う) | to nurture, feed |

泳 栄
害 照
障 清
阻 培
融 養

# 泳

| | | |
|---|---|---|
| 交差免疫電気泳動 | コウサメンエキデンキエイドウ | crossed immumo-electrophoresis |
| 細胞電気泳動 | サイボウデンキエイドウ | cell electrophoresis |
| 双頭交差免疫電気泳動 | ソウトウコウサメンエキデンキエイドウ | tandem crossed immuno-electrophoresis |
| 直線免疫電気泳動 | チョクセンメンエキデンキエイドウ | line immunoelectro-phoresis |
| 定量免疫電気泳動 | テイリョウメンエキデンキエイドウ | quantitative immuno-electrophoresis |
| 電気泳動 ★ | デンキエイドウ | electrophoresis |
| 電気泳動移動度 | デンキエイドウイドウド | electrophoretic mobility |
| 電気泳動注入 | デンキエイドウチュウニュウ | electrophoretic injection |
| デンプンゲル電気泳動 | デンプンゲルデンキエイドウ | starch gel electrophoresis |
| 等速電気泳動 | トウソクデンキエイドウ | isotachophoresis |
| 等電点電気泳動 ★ | トウデンテンデンキエイドウ | isoelectric focusing |
| 二次元電気泳動 | ニジゲンデンキエイドウ | two-dimensional electro-phoresis |
| 免疫電気泳動 ★ | メンエキデンキエイドウ | immunoelectrophoresis |
| 沪紙電気泳動 | ロシデンキエイドウ | paper electrophoresis |

# 栄

| | | |
|---|---|---|
| 栄養 ★ | エイヨウ | nutrition |
| 栄養生殖 ★ | エイヨウセイショク | vegetative reproduction |
| 栄養素 ★ | エイヨウソ | nutrient |
| 栄養繁殖 | エイヨウハンショク | vegetative propagation |
| 栄養ブイヨン | エイヨウブイヨン | nutrient broth |
| 栄養補給 | エイヨウホキュウ | alimentation |
| 寄生栄養 | キセイエイヨウ | paratrophic |
| 原栄養体 | ゲンエイヨウタイ | prototroph |
| 合胞体栄養細胞 | ゴウホウタイエイヨウサイボウ | syncytiotrophoblast |
| 光無機栄養 | コウムキエイヨウ | photolithotrophy |
| 光無機栄養生物 | コウムキエイヨウセイブツ | photolithotroph |
| 光有機栄養 | コウユウキエイヨウ | photoorganotrophy |
| 光有機栄養生物 | コウユウキエイヨウセイブツ | photoorganotroph |
| 混合栄養 | コンゴウエイヨウ | mixotrophism |
| 絶対光栄養生物 | ゼッタイヒカリエイヨウセイブツ | obligate phototroph |
| 絶対独立栄養生物 | ゼッタイドクリツエイヨウセイブツ | obligate autotroph |
| 蛋白質栄養障害 | タンパクシツエイヨウショウガイ | protein malnutrition |
| 独立栄養 | ドクリツエイヨウ | autotrophy |
| 独立栄養体 | ドクリツエイヨウタイ | autotroph |

| | | |
|---|---|---|
| 任意独立栄養生物 | ニンイドクリツエイヨウセイブツ | facultative autotroph |
| 光栄養生物 | ヒカリエイヨウセイブツ | phototroph |
| 光独立栄養生物 | ヒカリドクリツエイヨウセイブツ | photoautotroph |
| 無機栄養素 | ムキエイヨウソ | mineral nutrient |
| 無機栄養生物 | ムキエイヨウセイブツ | lithotroph |
| 有機栄養生物 | ユウキエイヨウセイブツ | organotroph |

## 害

| | | |
|---|---|---|
| 可逆阻害 | カギャクソガイ | reversible inhibition |
| 酵素阻害物質 | コウソソガイブッシツ | enzyme inhibitor |
| 細胞障害 | サイボウショウガイ | cell injury, cell damage |
| 弱有害遺伝子 | ジャクユウガイイデンシ | mildly deleterious gene |
| 障害 ★ | ショウガイ | damage, injury |
| 傷害 | ショウガイ | injury |
| 障害遺伝子 | ショウガイイデンシ | detrimental gene |
| 生物災害 | セイブツサイガイ | biohazard |
| 相関阻害 | ソウカンソガイ | coordinate repression |
| 阻害 ★ | ソガイ | inhibition |
| 阻害剤 | ソガイザイ | inhibitor, blocking reagent |
| 阻害蛋白質 | ソガイタンパクシツ | antizyme |
| 阻害定数 | ソガイテイスウ | inhibition constant |
| 阻害ペプチド | ソガイペプチド | inhibitor peptide |
| 蛋白質栄養障害 | タンパクシツエイヨウショウガイ | protein malnutrition |
| 放射線障害 | ホウシャセンショウガイ | radiation damage, radiation hazard |
| 有害 ★ | ユウガイ | harmful, hazardous, toxic |
| 有害種 | ユウガイシュ | harmful species |
| 累積阻害 | ルイセキソガイ | cumulative inhibition |

## 照

| | | |
|---|---|---|
| 照射 ★ | ショウシャ | irradiation |
| 照射線量 | ショウシャセンリョウ | exposure dose |
| 正の対照 | セイのタイショウ | positive control |
| 対照 ★ | タイショウ | control |
| 対照細胞培養 | タイショウサイボウバイヨウ | reference cell culture |
| 対照実験 | タイショウジッケン | control experiment |
| 対照テスト | タイショウテスト | control test |
| 負の対照 | フのタイショウ | negative control |

## 障

| | | |
|---|---|---|
| 細胞障害 ★ | サイボウショウガイ | cell injury |
| 障害 ★ | ショウガイ | damage, injury |
| 障害遺伝子 ★ | ショウガイイデンシ | detrimental gene |

| | | |
|---|---|---|
| 障壁 | ショウヘキ | barrier |
| 蛋白質栄養障害 | タンパクシツエイヨウショウガイ | protein malnutrition |
| 白内障 | ハクナイショウ | cataract |
| 放射線障害 | ホウシャセンショウガイ | radiation damage, radiation hazard |
| 緑内障 | リョクナイショウ | glaucoma |

## 清

| | | |
|---|---|---|
| 牛血清 | うしケッセイ | bovine serum |
| 牛胎児血清 | うしタイジケッセイ | fetal bovine serum |
| 血清 ★ | ケッセイ | (blood) serum |
| 血清学 | ケッセイガク | serology |
| 血清学的診断 | ケッセイガクテキシンダン | serodiagnosis |
| 血清型 | ケッセイがた | serotype |
| 血清反応 | ケッセイハンノウ | serological reaction |
| 血清療法 | ケッセイリョウホウ | serotherapy |
| 抗血清 ★ | コウケッセイ | antiserum |
| 子牛血清 | こうしケッセイ | calf serum |
| 高度免疫血清 | コウドメンエキケッセイ | hyperimmune serum |
| 上清 ★ | ジョウセイ | supernatant |
| 清澄因子リパーゼ | セイチョウインシリパーゼ | clearing factor lipase |
| 清澄化因子 | セイチョウカインシ | clearing factor |
| 多価抗血清 | タカコウケッセイ | polyvalent antiserum |
| 単一特異性抗血清 | タンイツトクイセイコウケッセイ | monospecific antiserum |
| 無血清培地 | ムケッセイバイチ | serum-free medium |
| 免疫血清 | メンエキケッセイ | immune serum |

## 阻

| | | |
|---|---|---|
| 可逆阻害 | カギャクソガイ | reversible inhibition |
| 酵素阻害物質 | コウソソガイブッシツ | enzyme inhibitor |
| 相関阻害 | ソウカンソガイ | coordinate repression |
| 阻害 ★ | ソガイ | inhibition |
| 阻害剤 ★ | ソガイザイ | inhibitor, blocking reagent |
| 阻害蛋白質 ★ | ソガイタンパクシツ | antizyme |
| 阻害定数 | ソガイテイスウ | inhibition constant |
| 阻害ペプチド | ソガイペプチド | inhibitor peptide |
| 累積阻害 | ルイセキソガイ | cumulative inhibition |

## 培

| | | |
|---|---|---|
| 液体培地 | エキタイバイチ | liquid medium |
| 回転培養 | カイテンバイヨウ | rotation culture |
| かくはん培養 | かくはんバイヨウ | spinner culture |
| 機能培養 | キノウバイヨウ | functional culture |

| | | |
|---|---|---|
| 継代培養 | ケイダイバイヨウ | subculture |
| 合成培地 | ゴウセイバイチ | synthetic medium |
| 固形培地 | コケイバイチ | solid medium |
| 固形培養 | コケイバイヨウ | solid culture |
| 最少培地 | サイショウバイチ | minimal medium |
| 細胞培養 | サイボウバイヨウ | cell culture |
| 斜面培養 | シャメンバイヨウ | slant culture |
| 集団培養 | シュウダンバイヨウ | mass culture |
| 純培養 | ジュンバイヨウ | pure culture |
| 振とう培養 | シントウバイヨウ | shake culture |
| 増殖培地 | ゾウショクバイチ | growth medium |
| 組織培養 ★ | ソシキバイヨウ | tissue culture |
| 対照細胞培養 | タイショウサイボウバイヨウ | reference cell culture |
| 大量培養 | タイリョウバイヨウ | mass culture |
| 単細胞培養 | タンサイボウバイヨウ | single cell culture |
| 短期培養 | タンキバイヨウ | short-term culture |
| 同調培養 | ドウチョウバイヨウ | synchronous culture |
| 培地 ★ | バイチ | (culture) medium |
| 培養 ★ | バイヨウ | culture |
| 培養液 | バイヨウエキ | (culture) medium |
| 培養細胞 | バイヨウサイボウ | cultured cell |
| バッチ式培養 | バッチシキバイヨウ | batch culture |
| 非同調培養 | ヒドウチョウバイヨウ | random culture, nonsynchronous culture |
| 無血清培地 | ムケッセイバイチ | serum-free medium |
| 連続培養 | レンゾクバイヨウ | continuous culture |

## 融

| | | |
|---|---|---|
| 遺伝子融合 ★ | イデンシユウゴウ | gene fusion |
| 核融合 | カクユウゴウ | nuclear fusion, karyogamy |
| 細胞融合 ★ | サイボウユウゴウ | cell fusion |
| 中心粒融合 | チュウシンリュウユウゴウ | centric fusion, centriole fusion |
| 動原体融合 | ドウゲンタイユウゴウ | centric fusion, centromere fusion |
| 融解 ★ | ユウカイ | fusion |
| 融解温度 | ユウカイオンド | melting temperature |
| 融合 | ユウゴウ | fusion |
| 融合遺伝 | ユウゴウイデン | blending inheritance |
| 融合核 | ユウゴウカク | synkaryon |
| 融合細胞 | ユウゴウサイボウ | fused cell, syncytium |
| 融点 | ユウテン | melting point |

## 養

| | | |
|---|---|---|
| 栄養 ★ | エイヨウ | nutrition |
| 栄養生殖 | エイヨウセイショク | vegetative reproduction |
| 栄養素 | エイヨウソ | nutrient |
| 栄養繁殖 | エイヨウハンショク | vegetative propagation |
| 栄養ブイヨン | エイヨウブイヨン | nutrient broth |
| 回転培養 | カイテンバイヨウ | rotation culture |
| かくはん培養 | かくはんバイヨウ | spinner culture |
| 機能培養 | キノウバイヨウ | functional culture |
| 継代培養 | ケイダイバイヨウ | subculture |
| 固形培養 | コケイバイヨウ | solid culture |
| 細胞培養 | サイボウバイヨウ | cell culture |
| 斜面培養 | シャメンバイヨウ | slant culture |
| 集団培養 | シュウダンバイヨウ | mass culture |
| 純培養 | ジュンバイヨウ | pure culture |
| 振とう培養 | シントウバイヨウ | shake culture |
| 組織培養 ★ | ソシキバイヨウ | tissue culture |
| 対照細胞培養 | タイショウサイボウバイヨウ | reference cell culture |
| 大量培養 | タイリョウバイヨウ | mass culture |
| 短期培養 | タンキバイヨウ | short-term culture |
| 単細胞培養 | タンサイボウバイヨウ | single cell culture |
| 蛋白質栄養障害 | タンパクシツエイヨウショウガイ | protein malnutrition |
| 同調培養 | ドウチョウバイヨウ | synchronous culture |
| 培養 ★ | バイヨウ | culture |
| 培養液 | バイヨウエキ | (culture) medium |
| 培養細胞 | バイヨウサイボウ | cultured cell |
| バッチ式培養 | バッチシキバイヨウ | batch culture |
| 非同調培養 | ヒドウチョウバイヨウ | random culture |
| 連続培養 | レンゾクバイヨウ | continuous culture |

*SUPPLEMENTARY VOCABULARY USING KANJI FROM CHAPTER 12*

| | | |
|---|---|---|
| 遺伝 | イデン | inheritance |
| 遺伝学 | イデンガク | genetics |
| 遺伝形質 | イデンケイシツ | genetic trait, inherited character |
| 遺伝子 | イデンシ | gene |
| 遺伝子組変え | イデンシくみかえ | genetic recombination |
| 遺伝子操作 | イデンシソウサ | gene manipulation |
| 遺伝子増幅 | イデンシゾウフク | gene amplification |
| 移動度 | イドウド | mobility |

| | | |
|---|---|---|
| 核 | カク | nucleus |
| 核移植 | カクイショク | nuclear transplantation |
| 核移入 | カクイニュウ | nucleus introduction |
| 核酸 | カクサン | nucleic acid |
| 核酸分解酵素 | カクサンブンカイコウソ | nuclease |
| 核分裂 | カクブンレツ | nuclear division |
| 核膜 | カクマク | nuclear membrane |
| 核膜孔 | カクマクコウ | nuclear pore |
| 重なり遺伝子 | かさなりイデンシ | overlapping gene |
| 滑面小胞体 | カツメンショウホウタイ | smooth(-surfaced) endoplasmic reticulum |
| 過敏性 | カビンセイ | hypersensitivity |
| 偽遺伝子 | ギイデンシ | pseudogene |
| 機構 | キコウ | mechanism |
| 組込み | くみこみ | integration |
| 形質膜 | ケイシツマク | plasma membrane |
| 形成 | ケイセイ | formation |
| 形態形成 | ケイタイケイセイ | morphogenesis |
| 血球 | ケッキュウ | hemocyte, blood cell |
| 原核細胞 | ゲンカクサイボウ | procaryotic cell |
| 原核生物 | ゲンカクセイブツ | procaryote |
| 原形質 | ゲンケイシツ | protoplasm |
| 原腸形成 | ゲンチョウケイセイ | gastrulation |
| 細胞 | サイボウ | cell |
| 細胞遺伝学 | サイボウイデンガク | cytogenetics |
| 細胞外酵素 | サイボウガイコウソ | exoenzyme, extracellular enzyme |
| 細胞学 | サイボウガク | cytology |
| 細胞間質 | サイボウカンシツ | intercellular substance |
| 細胞間マトリックス | サイボウカンマトリックス | intercellular matrix |
| 細胞質 | サイボウシツ | cytoplasm |
| 細胞周期 | サイボウシュウキ | cell cycle |
| 細胞小器官 | サイボウショウキカン | organelle |
| 細胞生物学 | サイボウセイブツガク | cell biology |
| 細胞増殖巣 | サイボウゾウショクソウ | focus |
| 細胞分裂 | サイボウブンレツ | cell division |
| 細胞膜 | サイボウマク | cytoplasmic membrane, cell membrane |
| 作動遺伝子 | サドウイデンシ | operator gene |
| 子嚢胞子 | シノウホウシ | ascospore |
| 従性遺伝 | ジュウセイイデン | sex-controlled inheritance |
| 消光 | ショウコウ | quenching |
| 小胞 | ショウホウ | vesicle |

| | | |
|---|---|---|
| 小胞体 | ショウホウタイ | endoplasmic reticulum |
| 植物 | ショクブツ | plant |
| 生殖細胞 | セイショクサイボウ | germ cell |
| 赤血球 | セッケッキュウ | erythrocyte, red blood cell |
| 染色体粒 | センショクタイリュウ | nucleosome |
| 相似形質 | ソウジケイシツ | analog |
| 組織 | ソシキ | tissue, texture, organization |
| 粗面小胞体 | ソメンショウホウタイ | rough(-surfaced) endoplasmic reticulum |
| 体細胞 | タイサイボウ | somatic cell |
| 多遺伝子 | タイデンシ | polygene |
| 多遺伝子遺伝 | タイデンシイデン | multigenic inheritance |
| 対立遺伝子 | タイリツイデンシ | allele |
| 単位胞 | タンイホウ | unit cell |
| 単核細胞 | タンカクサイボウ | mononuclear cell |
| 単細胞生物 | タンサイボウセイブツ | unicellular organism |
| 注射 | チュウシャ | injection |
| 中心粒 | チュウシンリュウ | centriole |
| 電気的細胞計数器 | デンキテキサイボウケイスウキ | electronic cell counter |
| 伝令RNA | デンレイRNA | messenger RNA |
| 透過性 | トウカセイ | permeability |
| 同義遺伝子 | ドウギイデンシ | multiple genes, polymeric gene |
| 白血球 | ハッケッキュウ | leukocyte, white blood cell |
| 伴性遺伝 | ハンセイイデン | sex-linked inheritance |
| 封入体細胞 | フウニュウタイサイボウ | inclusion cell |
| 分子遺伝学 | ブンシイデンガク | molecular genetics |
| 分子細胞遺伝学 | ブンシサイボウイデンガク | molecular cytogenetics |
| 胞子 | ホウシ | spore |
| 胞子形成 | ホウシケイセイ | sporulation |
| 放射 | ホウシャ | radiation |
| 放射性廃棄物 | ホウシャセイハイキブツ | radioactive waste |
| 放射線 | ホウシャセン | radiation |
| 無核細胞 | ムカクサイボウ | akaryote |
| 娘細胞 | むすめサイボウ | daughter cell |
| 無担体放射性同位体 | ムタンタイホウシャセイドウイタイ | carrier-free radioisotope |
| 卵形成 | ランケイセイ | oogenesis |
| 卵胞 | ランホウ | (ovarian) follicle |
| リンパ球 | リンパキュウ | lymph cell, lymphocyte |
| 沪過器 | ロカキ | filter |
| 沪過滅菌 | ロカメッキン | sterilization by filtration |

# EXERCISES

### Ex. 2.1 Matching Japanese and English terms

( ) 遺伝子融合
( ) 栄養ブイヨン
( ) 液体培地
( ) 酵素阻害物質
( ) 障害遺伝子

( ) 振とう培養
( ) 増殖培地
( ) 阻害蛋白質
( ) 対照細胞培養

( ) 蛋白質栄養障害
( ) デンプンゲル電気泳動
( ) 無血清培地
( ) 免疫電気泳動

1. antizyme
2. detrimental gene
3. enzyme inhibitor
4. gene fusion
5. growth medium
6. immunoelectrophoresis
7. liquid medium
8. nutrient broth
9. protein malnutrition
10. reference cell culture
11. serum-free medium
12. shake culture
13. starch gel electrophoresis

### Ex. 2.2 KANJI with similar structural elements

Look carefully at each of the two KANJI on the left, and note which structural element is common to both. Combine each KANJI on the left with the appropriate KANJI on the right to make a meaningful JUKUGO. Each technical term that contains one or more of the 100 KANJI introduced in this book can be found in the vocabulary lists for those KANJI. Other terms can be found in one of the supplementary vocabulary lists, including Lesson 0.

| | | | | |
|---|---|---|---|---|
| 1. | (1) 泳 | (2) 液 | 電気( )動 | 培養( ) |
| 2. | (1) 栄 | (2) 学 | 分子遺伝( ) | ( )養素 |
| 3. | (1) 泳 | (2) 清 | 細胞電気( )動 | 免疫血( ) |
| 4. | (1) 障 | (2) 降 | 沈( ) | ( )壁 |
| 5. | (1) 障 | (2) 阻 | ( )害定数 | 放射線( )害 |
| 6. | (1) 照 | (2) 熱 | ( )射線量 | ( )ショック蛋白質 |
| 7. | (1) 阻 | (2) 組 | ( )害ペプチド | ( )織培養 |
| 8. | (1) 培 | (2) 塩 | 脱( ) | ( )地 |
| 9. | (1) 融 | (2) 触 | 細胞( )合 | ( )媒還元 |
| 10. | (1) 養 | (2) 着 | 栄( ) | 脱( ) |

**Ex. 2.3 Matching Japanese technical terms with definitions**

Read each definition carefully, and then choose the appropriate technical term. Words that you have not yet encountered are listed following the definitions.

| | | |
|---|---|---|
| (　) 栄養生殖 | (　) 細胞障害 | (　) 照射 |
| (　) 血清 | (　) 細胞融合 | (　) 組織培養 |
| (　) 酵素阻害物質 | (　) 弱有害遺伝子 | (　) 免疫電気泳動 |
| (　) 最少培地 | | |

1. 電気泳動と組み合わせた、抗原抗体反応による検出定量方法。
2. 植物が胞子によって増殖したり、単細胞生物が分裂によって増殖する場合などのように細胞単位で増えること。
3. 酵素のある特定の部位に結合して反応速度を低下させる物質。
4. 何らかの形で、細胞に病理的な変化をもたらすこと。
5. ホモ接合体の生存力を2～3%程度減少させる突然変異を起こす遺伝子。
6. 物質を$\gamma$線、中性子線、電子線などにさらすこと。
7. 凝固していない血液から血球を除いたもの。
8. 生育に必要最少限の成分のみからなる培地。
9. 生体の組織の一部を取り出し、その特性を保持しつつ培養すること。
10. 2個以上の細胞が融合し、単一の細胞膜で包まれた細胞が生ずる現象。

| | | | | | |
|---|---|---|---|---|---|
| 検出 | ケンシュツ | detection | 凝固 | ギョウコ | coagulation |
| 特定 | トクテイ | specific | 除く | のぞく | to remove |
| 低下 | テイカ | decrease | 生育 | セイイク | growth |
| 何らかの | なんらかの | some kind of | 最少限 | サイショウゲン | minimum |
| 病理的 | ビョウリテキ | pathological | 取り出す | とりだす | to take (out) |
| ホモ 接合体 | ホモ セツゴウタイ | homozygote | 特性 | トクセイ | characteristics |
| | | | 保持 | ホジ | preservation |
| 生存力 | セイゾンリョク | viability | 2個 | ニコ | two (of something) |
| 突然変異 | トツゼンヘンイ | mutation | | | |
| さらす | | to expose | 包む | つつむ | to envelop |

**Ex. 2.4 Sentence translations**

Read each sentence carefully, and then translate it. Words that you have not yet encountered are listed following the sentences.

1. ゲルは分子ふるいとして機能するので、同一の電気泳動移動度をもつ物質でも、大きさや形状を異にするならば分離することができる。
2. 等電点を異にする種々の両性電解質を低イオン強度下で電気泳動すると、各々の両性電解質は等電点の順序に配列するように泳動し、この配列が達成されると静止し、相互分離する。
3. 独立栄養の高等植物にとっての栄養素は、窒素、カリウム、リンが基本であり、これを植物の三大栄養素という。
4. 電離性放射線の作用により、生体の細胞や組織が変化し、細胞の分裂阻害・突然変異・死滅、組織の破壊などによって、人体が受ける障害を放射線障害という。
5. 累積阻害の場合は代謝経路での最終産物が複数個存在し、各産物による阻害が各々の部分的でかつ互いに独立して作用する。
6. 生物を対象とするような複雑な系では、ある実験操作は複数の要素的条件を含むのが普通なので、対照実験は必ず必要とされる。
7. 動物細胞の培地として血清の代わりにホルモン、結合蛋白質、細胞接着因子などを添加すれば、細胞の生存や増殖ができるが、無血清培地の培養における物理化学的環境の最適条件は血清培地に比べて厳密である。
8. 固形培養、つまり微生物や動植物の細胞の固形培地での培養を行う場合、通常寒天やゼラチンを用いるが、シリカゲル、パンなどを使用することも多い。
9. 大量培養を大別すれば大型の培養容器を用いて大量に培養し一度に収穫するバッチ式培養と、培地を補給しながら繰返し収穫する連続培養とがある。
10. 組換えDNA手法により、調べたい遺伝子に大腸菌の*lacZ*遺伝子を人工的に融合し、β-カラクトシダーゼ活性を標識にして、遺伝子表現の様相や遺伝子産物の細胞内分布を調べる方法が広くとられている。

| | | |
|---|---|---|
| 分子ふるい | ブンシふるい | molecular sieve |
| 形状 | ケイジョウ | shape |
| 異にする | ことにする | to differ |
| 両性 | リョウセイ | amphoteric |
| 低- | テイ- | low |
| 各々の | おのおのの | each |
| 順序 | ジュンジョ | sequence, order |
| 達成 | タッセイ | attainment |
| 静止 | セイシ | stillness |
| 相互- | ソウゴ | mutual |
| 窒素 | チッソ | nitrogen |
| カリウム | | potassium |
| リン | | phosphorus |
| 基本 | キホン | foundation |
| 電離性 | デンリセイ | ionizing |
| 死滅 | シメツ | death, extinction |
| 人体 | ジンタイ | human body |
| 受ける | うける | to receive |
| 代謝 | タイシャ | metabolism |
| 経路 | ケイロ | pathway, route |
| 最終 | サイシュウ | final |
| 産物 | サンブツ | product |
| 複数個 | フクスウコ | several (of something) |
| 各 | カク | each |
| 互いに | たがいに | mutually |
| 独立 | ドクリツ | independence |
| 複雑な | フクザツな | complex |

| | | | | | |
|---|---|---|---|---|---|
| 操作 | ソウサ | operation, manipulation | 容器 | ヨウキ | vessel, container |
| 要素的 | ヨウソテキ | essential | 収穫 | シュウカク | harvest |
| 普通 | フツウ | usual, common | 繰返す | くりかえす | to repeat |
| 添加 | テンカ | addition | 組換え | くみかえ | recombination |
| 生存 | セイゾン | existence | 手法 | シュホウ | technique |
| 最適 | サイテキ | optimum | 調べる | しらべる | to investigate |
| 通常 | ツウジョウ | ordinarily | 人工的 | ジンコウテキ | artificial |
| パン | | bread | 標識 | ヒョウシキ | label |
| 大別 | タイベツ | broad classification | 表現 | ヒョウゲン | expression |
| 大型 | おおがた | large (size) | 様相 | ヨウソウ | aspect, condition |
| | | | 分布 | ブンプ | distribution |

**Ex. 2.5 Additional dictionary entries**

( ) 偽遺伝子　( ) 赤血球　( ) 対立遺伝子
( ) 細胞遺伝学　( ) 体細胞　( ) 白血球
( ) 細胞生物学　( ) 多遺伝子　( ) 分子遺伝学
( ) 作動遺伝子

1. 遺伝現象の機構を分子レベルで解析しようとする分野で、発生・分化・進化・老化・免疫などの解析に発展している学問。
2. 塩基配列において既知の遺伝子と明らかに相同であるが、遺伝子としては機能を失ったDNAの領域。
3. 主に染色体の形態、構造および行動などの細胞学的な特徴から遺伝現象を明らかにしようとする遺伝学の一分野。
4. 血液中の有形成分の大部分を占める、核をもたなくてもATP合成を行う細胞部分。
5. 個々の遺伝子の作用はきわめて弱いが、多数が同義的に補足しあいながらある量的形質の発現に関与する遺伝子群。
6. 生体物質とその代謝についての知識と細胞構造についての知識とを総合して、細胞レベルでさまざまな生命現象を研究しようとする学問。
7. 多細胞生物を構成している全細胞のうち生殖細胞以外のもの。
8. 同一遺伝子座に属し、互いに区別される遺伝的変異体。
9. 特異的なリプレッサーと直接結合し、そのオペロンに属する構造遺伝子の情報発現を開閉するスイッチの役割をもち、染色体上のいくつかのヌクレオチド対をもつ機能部分。
10. 哺乳動物の血球成分の一つで、核を備えた、形態的、機能的に異なったいくつかの細胞種の集合。

| | | | | | |
|---|---|---|---|---|---|
| 解析 | カイセキ | analysis | 発展 | ハッテン | development, growth |
| 進化 | シンカ | evolution | | | |
| 老化 | ロウカ | aging | | | |

| | | |
|---|---|---|
| 既知 | キチ | established, already known |
| 明らか | あきらか | clear, obvious |
| 失う | うしなう | to lose |
| 領域 | リョウイキ | region |
| 主に | おもに | principally |
| 特徴 | トクチョウ | characteristic |
| 有形 | ユウケイ | substantial |
| 個々の | ココの | individual |
| 弱い | よわい | weak |
| 同義的 | ドウギテキ | synonymous |
| 形質 | ケイシツ | character, trait, feature |
| 関与 | カンヨ | participation |
| -群 | -グン | group |

| | | |
|---|---|---|
| 総合する | ソウゴウする | to integrate, to put together |
| 遺伝子座 | イデンシザ | locus |
| 属する | ゾクする | to belong to |
| 変異体 | ヘンイタイ | mutant |
| 開閉 | カイヘイ | opening and shutting |
| 役割 | ヤクわり | role |
| 哺乳動物 | ホニュウドウブツ | mammal |
| 備える | そなえる | to be equipped with |
| 集合 | シュウゴウ | aggregation, assembly |

**Translations for Ex. 2.4**

1. A gel functions as a molecular sieve. Thus, even if substances possess identical electrophoretic mobility, if they differ in size or shape it is possible to separate them.
2. If we carry out under low ionic strength the electrophoresis of various amphoteric electrolytes with differing equivalence points, each amphoteric electrolyte will migrate to form a line in order of the equivalence points. Once this sequence has been attained, the electrolytes cease moving, and they have been mutually separated.
3. Nitrogen, potassium and phosphorus are the basic nutrients for autotrophic, higher order plants. These are called the three major nutrients for plants.
4. The cells and tissue of an organism change due to the action of ionizing radiation. The damage sustained by the human body resulting from the death, mutation, or inhibition of mitosis in cells, or from the breakdown of tissue, is called radiation hazard.
5. In cumulative inhibition several final products exist from a metabolic pathway. The inhibitions resulting from each product act separately and in a mutually independent way.
6. For a complex system in which we take living things to be the subject, it is common to include several essential conditions in a certain experimental operation. For this reason a control experiment is always necessary.
7. In place of (blood) serum as the medium for animal cells, if we add materials such as hormones, binding proteins or cell adhesion factors, it is possible for cells to exist and to grow. However, in cultures with serum-free medium the optimal conditions for the physicochemical environment are more strict than with medium containing serum.
8. When we carry out a solid culture, that is, the culture of microorganisms or of plant or animal cells on a solid medium, we generally use agar or gelatin. However, silica gel, bread, and other media are also used often.
9. If we classify mass culture broadly, we have batch culture—in which we employ a large culture vessel, culture a large quantity [of material], and harvest once—and we have continuous culture—in which we harvest repeatedly while replenishing the medium.
10. A method that has been widely adopted involves artifically fusing the gene we wish to investigate into the *lacZ* gene of *E. coli* using recombinant DNA techniques, taking β-galactosidase activity as a label, and investigating the condition of gene expression and the distribution of gene products within the cells.

# LESSON 3 第三課 BTJ 13

| | | |
|---|---|---|
| 寒 | カン | cold |
| | さむ(い) | cold, chilly |
| 嫌 | ケン | dislike, abhorrence |
| | いや(がる) | to dislike, abhor {v.t.} |
| | きら(う) | to dislike, abhor {v.t.} |
| | きら(い) | dislike, abhorrence |
| 呼 | コ | calling; designation |
| | よ(ぶ) | to call, designate |
| 好 | コウ | liking, fondness; good |
| | この(む) | to like, be fond of {v.t.} |
| | す(く) | to like, be fond of {v.t.} |
| | す(き) | liking, fondness |
| 止 | シ | stopping |
| | と(める) | to stop {v.t.} |
| | と(まる) | to stop {v.i.} |
| 謝 | シャ | apology; thanks |
| 節 | セツ | moderation; section |
| | ふし | joint, knuckle, lump |
| 濁 | ダク | turbidity |
| | にご(す) | to make muddy |
| | にご(る) | to be muddy |
| 天 | テン | heaven, nature |
| 冷 | レイ | cold |
| | つめ(たい) | cold, cool |
| | ひ(やす) | to cool, refrigerate |
| | ひ(える) | to grow cold, cool off |

寒 嫌
呼 好
止 謝
節 濁
天 冷

# 寒

| | | |
|---|---|---|
| 栄養寒天培地 | エイヨウカンテンバイチ | nutrient agar medium |
| 寒天 ★ | カンテン | agar |
| 寒天ゲル電気泳動 | カンテンゲルデンキエイドウ | agar gel electrophoresis |
| 寒天藻 | カンテンソウ | agarophyte |
| 寒天培地 | カンテンバイチ | agar medium |
| 寒冷生物 ★ | カンレイセイブツ | psychrophile |
| 寒冷不溶性グロブリン | カンレイフヨウセイグロブリン | cold insoluble globulin |
| 耐寒性 ★ | タイカンセイ | cryotolerance |
| 軟寒天培養 | ナンカンテンバイヨウ | soft agar culture |

# 嫌

| | | |
|---|---|---|
| 嫌気性細菌 ★ | ケンキセイサイキン | anaerobic bacteria |
| 嫌気性生物 | ケンキセイセイブツ | anaerobe |
| 嫌気性発酵 ★ | ケンキセイハッコウ | anaerobic fermentation |
| 嫌気的呼吸 ★ | ケンキテキコキュウ | anaerobic respiration |
| 嫌気的条件 | ケンキテキジョウケン | anaerobic condition |
| 嫌気的代謝 | ケンキテキタイシャ | anaerobic metabolism |
| 絶対嫌気性細菌 | ゼッタイケンキセイサイキン | strictly anaerobic bacterium |
| 絶対嫌気性生物 | ゼッタイケンキセイセイブツ | strict anaerobe |
| 通性嫌気性細菌 | ツウセイケンキセイサイキン | facultative anaerobic bacterium |
| 通性嫌気性生物 | ツウセイケンキセイセイブツ | facultative anaerobe |

# 呼

| | | |
|---|---|---|
| 嫌気的呼吸 ★ | ケンキテキコキュウ | anaerobic respiration |
| 好気的呼吸 ★ | コウキテキコキュウ | aerobic respiration |
| 光呼吸 | コウコキュウ | photorespiration |
| 呼吸 ★ | コキュウ | respiration |
| 呼吸計 | コキュウケイ | respirometer |
| 呼吸酵素 | コキュウコウソ | respiratory enzyme |
| 呼吸色素 | コキュウシキソ | respiratory pigment |
| 呼吸商 | コキュウショウ | respiratory quotient |
| 呼吸調節 | コキュウチョウセツ | respiratory control |
| 状態四呼吸 | ジョウタイヨンコキュウ | state 4 respiration |
| 無機呼吸 | ムキコキュウ | inorganic respiration |

# 好

| | | |
|---|---|---|
| 好圧性 | コウアツセイ | barophilic |
| 好アルカリ性[細]菌 ★ | コウアルカリセイ[サイ]キン | alkalophilic bacterium |

| | | |
|---|---|---|
| 好塩基球 | コウエンキキュウ | basophil |
| 好塩菌 | コウエンキン | halophilic bacterium |
| 好塩性 | コウエンセイ | halophilic |
| 好塩性生物 | コウエンセイセイブツ | halophile |
| 好気性細菌 | コウキセイサイキン | aerobic bacterium |
| 好気性生物 | コウキセイセイブツ | aerobe |
| 好気的呼吸 ★ | コウキテキコキュウ | aerobic respiration |
| 好気的代謝 | コウキテキタイシャ | aerobic metabolism |
| 好酸球 | コウサンキュウ | eosinophil |
| 好酸性細菌 ★ | コウサンセイサイキン | acidophilic bacterium |
| 好酸性細胞 | コウサンセイサイボウ | acidophil |
| 好中球 | コウチュウキュウ | neutrophil |
| 好熱菌 | コウネツキン | thermophilic bacterium |
| 好熱性 | コウネツセイ | thermophilic |
| 好熱生物 | コウネツセイブツ | thermophile |
| 好冷菌 | コウレイキン | psychrophilic bacterium |
| 好冷性 | コウレイセイ | psychrophilic |
| 高度好熱菌 | コウドコウネツキン | extreme thermophile |
| 絶対好気性細菌 | ゼッタイコウキセイサイキン | strictly aerobic bacterium |
| 絶対好気性生物 | ゼッタイコウキセイセイブツ | strict aerobe |
| 絶対好熱菌 | ゼッタイコウネツキン | strict thermophile |
| 中等度好熱菌 | チュウトウドコウネツキン | moderate thermophile |
| 通性好熱菌 | ツウセイコウネツキン | facultative thermophile |

## 止

| | | |
|---|---|---|
| 休止 | キュウシ | diapause |
| 休止状態 ★ | キュウシジョウタイ | dormancy |
| 禁止クローン ★ | キンシクローン | forbidden clone |
| 迅速停止法 | ジンソクテイシホウ | rapid quenching method |
| 阻止 | ソシ | block |
| 阻止抗体 ★ | ソシコウタイ | blocking antibody |
| 沈降阻止反応 | チンコウソシハンノウ | precipitation inhibition reaction |
| 停止 | テイシ | stoppage, halt |
| 停止細胞 | テイシサイボウ | arrested cell, static cell |
| 凍結防止効果 | トウケツボウシコウカ | antifreeze effect |
| 防止 | ボウシ | prevention |
| 行き止まり阻害 | ゆきどまりソガイ | dead end inhibition |

## 謝

| | | |
|---|---|---|
| 一次代謝産物 | イチジタイシャサンブツ | primary metabolite |
| 基礎代謝 | キソタイシャ | basal metabolism |
| 嫌気的代謝 | ケンキテキタイシャ | anaerobic metabolism |

| | | |
|---|---|---|
| 好気的代謝 | コウキテキタイシャ | aerobic metabolism |
| 先天性代謝異常 | センテンセイタイシャイジョウ | inborn errors of metabolism |
| 代謝 ★ | タイシャ | metabolism |
| 代謝期 | タイシャキ | metabolic stage |
| 代謝経路 ★ | タイシャケイロ | metabolic pathway |
| 代謝性 | タイシャセイ | metabolic |
| 代謝制御 | タイシャセイギョ | metabolic control |
| 代謝生成物 | タイシャセイセイブツ | metabolite |
| 代謝阻害剤 | タイシャソガイザイ | metabolic inhibitor |
| 代謝速度 | タイシャソクド | metabolic rate |
| 代謝中間体 | タイシャチュウカンタイ | metabolic intermediate |
| 代謝調節 ★ | タイシャチョウセツ | metabolic regulation |
| 蛋白質代謝 | タンパクシツタイシャ | protein metabolism |
| 窒素代謝 | チッソタイシャ | nitrogen metabolism |
| 中間代謝 | チュウカンタイシャ | intermediary metabolism |
| 中間代謝物質 | チュウカンタイシャブッシツ | intermediary metabolite |
| 二次代謝産物 | ニジタイシャサンブツ | secondary metabolite |

## 節

| | | |
|---|---|---|
| 関節 | カンセツ | joint |
| 緊縮調節 | キンシュクチョウセツ | stringent control |
| 結節 | ケッセツ | node, nodule |
| 硬節 | コウセツ | sclerotome |
| 呼吸調節 ★ | コキュウチョウセツ | respiratory control |
| 自律神経節 | ジリツシンケイセツ | autonomic ganglion |
| 神経節 | シンケイセツ | ganglion |
| 正の調節 | セイのチョウセツ | positive regulation |
| 節足動物 | セッソクドウブツ | arthropod |
| 代謝調節 ★ | タイシャチョウセツ | metabolic regulation |
| 体節 | タイセツ | somite |
| 調節 | チョウセツ | control, regulation |
| 調節遺伝子 | チョウセツイデンシ | regulator gene |
| 調節酵素 ★ | チョウセツコウソ | regulatory enzyme |
| 調節サブユニット | チョウセツサブユニット | regulatory subunit |
| 調節蛋白質 | チョウセツタンパクシツ | regulatory protein |
| 調節部位 | チョウセツブイ | regulatory site |
| 頭部神経節 | トウブシンケイセツ | cephalic ganglion |
| 二次小結節 | ニジショウケッセツ | secondary nodule |
| 分節 | ブンセツ | segment |
| 翻訳後調節 | ホンヤクゴチョウセツ | posttranslational control |
| 翻訳調節 | ホンヤクチョウセツ | translational control |
| 慢性関節リウマチ | マンセイカンセツリウマチ | rheumatoid arthritis |

## 濁

| | | |
|---|---|---|
| 汚濁 ★ | オダク | pollution |
| 懸濁[液] ★ | ケンダク[エキ] | suspension |
| 懸濁培養 ★ | ケンダクバイヨウ | suspension culture |
| 濁度 | ダクド | turbidity |
| 濁りプラーク | にごりプラーク | turbid plaque |
| 濁り溶菌斑 | にごりヨウキンハン | turbid plaque |
| 乳濁液 | ニュウダクエキ | emulsion, latex |
| 比濁計 | ヒダクケイ | nephelometer |

## 天

| | | |
|---|---|---|
| 栄養寒天培地 | エイヨウカンテンバイチ | nutrient agar medium |
| 寒天 ★ | カンテン | agar |
| 寒天ゲル電気泳動 | カンテンゲルデンキエイドウ | agar gel electrophoresis |
| 寒天藻 | カンテンソウ | agarophyte |
| 寒天培地 | カンテンバイチ | agar medium |
| 後天性免疫 ★ | コウテンセイメンエキ | acquired immunity |
| 後天的 | コウテンテキ | acquired |
| 先天性代謝異常 | センテンセイタイシャイジョウ | inborn errors of metabolism |
| 先天性免疫 | センテンセイメンエキ | innate immunity |
| 先天的 | センテンテキ | natural, innate |
| 天然- ★ | テンネン- | natural |
| 天然酵素 | テンネンコウソ | natural enzyme |
| 天然培地 | テンネンバイチ | natural medium |
| 天秤 | テンビン | balance |
| 軟寒天培養 | ナンカンテンバイヨウ | soft agar culture |

## 冷

| | | |
|---|---|---|
| 寒冷生物 ★ | カンレイセイブツ | psychrophile |
| 寒冷不溶性グロブリン | カンレイフヨウセイグロブリン | cold insoluble globulin |
| 急冷 | キュウレイ | quenching |
| 好冷菌 ★ | コウレイキン | psychrophilic bacterium |
| 好冷性 | コウレイセイ | psychrophilic |
| 冷却 | レイキャク | cooling |
| 冷却遠心機 ★ | レイキャクエンシンキ | refrigerated centrifuge |
| 冷光 | レイコウ | luminescence |
| 冷蔵 | レイゾウ | cold storage, refrigeration |
| 冷凍 | レイトウ | freezing, refrigeration |

## *SUPPLEMENTARY VOCABULARY USING KANJI FROM CHAPTER 13*

| | | |
|---|---|---|
| 一遺伝子一酵素仮説 | イチイデンシイチコウソカセツ | one gene-one enzyme hypothesis |
| 一次構造 | イチジコウゾウ | primary structure |
| 遺伝子工学 | イデンシコウガク | genetic engineering |
| 塩基 | エンキ | base |
| 塩基対 | エンキツイ | base pair |
| 塩基の対合 | エンキのタイゴウ | base pairing |
| 基質 | キシツ | substrate |
| 基質結合部位 | キシツケツゴウブイ | substrate-binding site |
| 吸着 | キュウチャク | adsorption |
| 供与体 | キョウヨタイ | donor |
| 酵素工学 | コウソコウガク | enzyme engineering |
| 細胞工学 | サイボウコウガク | cell technology |
| 受精 | ジュセイ | fertilization |
| 常染色体 | ジョウセンショクタイ | autosome |
| 自律性 | ジリツセイ | autonomy |
| 人工酵素 | ジンコウコウソ | synthetic enzyme, synzyme |
| 生物工学 | セイブツコウガク | biotechnology, bioengineering |
| 世代 | セダイ | generation |
| 多糖 | タトウ | polysaccharide |
| 単糖 | タントウ | monosaccharide |
| 通気 | ツウキ | aeration |
| 電極 | デンキョク | electrode |
| 同調的 | ドウチョウテキ | coordinate, synchronous |
| 同調分裂 | ドウチョウブンレツ | synchronous division |
| 二極性 | ニキョクセイ | bipolar |
| 二次構造 | ニジコウゾウ | secondary structure |
| 乳糖 | ニュウトウ | milk sugar, lactose |
| 橋かけ構造 | はしかけコウゾウ | crosslinkage |
| ブドウ糖 | ブドウトウ | glucose, dextrose |
| 分極 | ブンキョク | polarization |
| 未受精卵 | ミジュセイラン | unfertilized egg |
| ゆらぎ塩基対 | ゆらぎエンキツイ | wobble base pair |
| ゆらぎ説 | ゆらぎセツ | wobble hypothesis |

## EXERCISES

### Ex. 3.1 Matching Japanese and English terms

( ) 寒天ゲル電気泳動
( ) 寒天培地
( ) 嫌気的呼吸
( ) 嫌気的代謝
( ) 懸濁培養

( ) 好気的呼吸
( ) 好気的代謝
( ) 好冷菌
( ) 呼吸調節

( ) 代謝調節
( ) 沈降阻止反応
( ) 軟寒天培養
( ) 冷却遠心機

1. aerobic metabolism
2. aerobic respiration
3. agar gel electrophoresis
4. agar medium
5. anaerobic metabolism
6. anaerobic respiration
7. metabolic regulation
8. precipitation inhibition reaction
9. psychrophilic bacterium
10. refrigerated centrifuge
11. respiratory control
12. soft agar culture
13. suspension culture

### Ex. 3.2 KANJI with the same ON reading

Look carefully at each of the two KANJI on the left, and note the ON reading that is common to both. Combine each KANJI on the left with the appropriate KANJI on the right to make a meaningful JUKUGO. Each technical term that contains one or more of the 100 KANJI introduced in this book can be found in the vocabulary lists for those KANJI. Other terms can be found in one of the supplementary vocabulary lists, including Lesson 0.

| | | | | |
|---|---|---|---|---|
| 1. | (1) 泳 | (2) 栄 | 電気( )動移動度 | ( )養生殖 |
| 2. | (1) 害 | (2) 外 | 細胞( )酵素 | 生物災( ) |
| 3. | (1) 嫌 | (2) 験 | ( )気的条件 | 対照実( ) |
| 4. | (1) 呼 | (2) 固 | ( )吸酵素 | ( )形培地 |
| 5. | (1) 好 | (2) 酵 | ( )アルカリ性細菌 | ( )素工学 |
| 6. | (1) 謝 | (2) 射 | 代( ) | 注( ) |
| 7. | (1) 照 | (2) 障 | ( )害 | ( )射 |
| 8. | (1) 清 | (2) 精 | 血( ) | 受( ) |
| 9. | (1) 天 | (2) 点 | 寒( ) | 融( ) |
| 10. | (1) 融 | (2) 有 | ( )合細胞 | 弱( )害遺伝子 |

**Ex. 3.3 Matching Japanese technical terms with definitions**

Read each definition carefully, and then choose the appropriate technical term. Words that you have not yet encountered are listed following the definitions.

| | | |
|---|---|---|
| (　) 寒天培地 | (　) 好気的代謝 | (　) 調節酵素 |
| (　) 嫌気性生物 | (　) 好冷菌 | (　) 停止細胞 |
| (　) 嫌気的呼吸 | (　) 呼吸酵素 | (　) 軟寒天培地 |
| (　) 懸濁培養 | | |

1. 栄養ブイヨンに寒天を1.2–1.5%添加してゲルにしたもので、細菌の分離、純培養、菌株保存などに用いられている基本培地の一つ。
2. 培地に0.3～0.6%程度の寒天を溶解固化したもの。
3. 酸素のある状態では生存が困難または不可能な生物。
4. 細胞中の最終電子受容体が分子状酸素以外の物質である場合。
5. 生物の呼吸に関係する酸化還元反応を促進する酵素の総称。
6. 分子状酸素を消費するエネルギー代謝の形式。
7. 一般には、20℃以下に最適温度をもつか、0℃近くの温度で生育できる細菌。
8. 細胞周期の進行をまったく止められた状態の細胞。
9. 代謝を調節する役割をもつ酵素。
10. 細胞を培養容器基質に付着させることなく、液体培地の中に浮遊させた状態で増殖させる培養。

| | | | | | |
|---|---|---|---|---|---|
| 菌株 | キンかぶ | strain | 消費 | ショウヒ | consumption |
| 保存 | ホゾン | conservation, stock | 進行 | シンコウ | progress |
| 困難 | コンナン | difficult | 付着 | フチャク | attachment |
| 不可能 | フカノウ | impossible | 浮遊 | フユウ | floating |

**Ex. 3.4 Sentence translations**

Read each sentence carefully, and then translate it. Words that you have not yet encountered are listed following the sentences.

1. 寒天、あるいは寒天の一成分であるアガロースは、加熱すると水に溶解する。これらの水溶液を40～50℃に保温し、これに生体触媒を加えて冷却すると固定化生体触媒が得られる。
2. グルコース1分子は好気的代謝ではクエン酸回路を通って、38分子のATPが生成されるが、嫌気的代謝である解糖やアルコール発酵では2分子のATPが生成されるにすぎない。
3. 絶対嫌気性細菌は酸素存在下で生育できない細菌で、有機基質の嫌気性発酵によりエネルギーを獲得する。
4. 進化的には、嫌気的呼吸が先に出現し、その後好気的呼吸が出現したと考えられている。エネルギーの獲得形式としては、嫌気的呼吸は好気的呼吸よりも効率が悪い。
5. 75℃以上でも生育できる細菌を高度好熱菌とよび、55℃から75℃まで生育できるものを中等度好熱菌とよぶ。

6. 従来、好冷生物としては、極地地方の海にすむ魚類がよく研究されており、例えば、魚は細胞の凍結を防止するため氷点降下剤として働くペプチドを生産することなどが知られている。
7. F. M. Burnetが抗体生産機構についてのクローン選択説において禁止クローンの存在を仮定した。正常動物において自己免疫が成立しないのは、禁止クローンが消滅しているためであると考えた。
8. 代謝は、隣接する個々の反応で成立しているが、その反応は個々の異なる酵素によって促進されている。したがって、代謝調節はそれらの酵素、特に律速段階の酵素の機能を変動させることで実現される。
9. 遺伝子発現はDNAからmRNAへの転写段階およびmRNAから蛋白質への翻訳段階の二つの段階で主に調節を受けていると考えられており、このうち後者の段階を翻訳調節という。
10. ビルレントファージが感受性菌の平板上で透明なプラークを形成することに対して、テンペレートファージは感受性菌をまいた平板上で濁りプラークを形成する。

| | | |
|---|---|---|
| 保温 | ホオン | keeping warm |
| 回路 | カイロ | cycle |
| 解糖 | カイトウ | glycolysis |
| 獲得 | カクトク | gain, aquisition |
| 従来 | ジュウライ | heretofore |
| 海 | うみ | sea |
| すむ | | to live |
| 魚類 | ギョルイ | fish (as a group) |
| 魚 | さかな | fish |
| 凍結 | トウケツ | freezing |
| 働く | はたらく | to act, to work |
| 選択 | センタク | selection |
| 仮定 | カテイ | supposition, hypothesis |
| 正常 | セイジョウ | normality |
| 消滅 | ショウメツ | disappearance |
| 律速 | リッソク | rate determining |
| 段階 | ダンカイ | step, stage |
| 実現 | ジツゲン | realization |
| 翻訳 | ホンヤク | translation |
| 後者 | コウシャ | the latter |
| 感受性菌 | カンジュセイキン | sensitive bacterium |
| 平板 | ヘイバン | flat plate |
| 透明 | トウメイ | transparency |
| まく | | to sow, to seed, to sprinkle |

**Ex. 3.5 Additional dictionary entries**

( ) 一遺伝子一酵素仮説
( ) 遺伝子工学
( ) 塩基対
( ) 酵素工学
( ) 細胞工学
( ) 受精
( ) 常染色体
( ) 人工酵素
( ) 同調分裂
( ) ゆらぎ説

1. 遺伝子が特定の一つの酵素の生成に関与し、その特異な作用を介して生物の表現型を支配しているという考え。
2. 核酸の塩基のうち定まった組合せの2個が互いに水素結合により対合したもの。
3. 組換えDNAやDNAクローニングなどの遺伝子操作を利用して有用物質を大量に生産しようとする学問分野。

4. 高選択性、高効率、活性調節機構などの天然酵素の機能をもつ人工触媒。
5. 酵素の産業的応用、つまり有用物質の生産、エネルギー関連物質の生産、環境保全、新しい医療への応用など、を目的とする工学。
6. コドンとアンチコドンが対合する際、コドンの3番目の塩基とアンチコドンの最初の塩基との対合にはゆらぎがあり、ワトソン-クリックの塩基対以外にいくつかの対合が可能とする考え方。
7. 細胞集団のすべての細胞が相のそろった細胞周期のリズムで細胞分裂を繰返すこと。
8. 細胞融合、細胞培養など細胞生物学上の手法を利用して、有用物質の生産や新製品種の人工育成をする技術。
9. 雌性配偶子である卵に、雄性配偶子である精子が侵入して両者が融合し、新個体の発生が開始されること。
10. 性決定に関与する染色体と違い、形の等しい相同対から成り、減数分裂時には規則正しい対合および分裂の作動をとる染色体。

| | | |
|---|---|---|
| -を介して | -をカイして | through |
| 表現型 | ヒョウゲンがた | phenotype |
| 利用 | リヨウ | utilization |
| 作製 | サクセイ | preparation |
| 支配 | シハイ | control, domination |
| 関連 | カンレン | connection, relation |
| 保全 | ホゼン | preservation, conservation |
| 医療 | イリョウ | medical treatment |
| 3番目 | サンバンめ | the third (one) |
| 最初 | サイショ | initial |
| ゆらぎ | | fluctuation, wobbling |
| 可能 | カノウ | possibility |
| 細胞集団 | サイボウ シュウダン | cell population |
| 製品 | セイヒン | manufactured goods; products |
| 育成 | イクセイ | rearing, raising |
| 技術 | ギジュツ | technology |
| 雌性 | シセイ | female |
| 配偶子 | ハイグウシ | gamete |
| 雄性 | ユウセイ | male |
| 侵入 | シンニュウ | invasion, |
| 両者 | リョウシャ | both |
| 開始 | カイシ | initiation, start |
| 減数分裂 | ゲンスウ ブンレツ | meiosis |
| 規則 | キソク | regular, systematic |
| 正しい | ただしい | |

**Ex. 3.6 Additional sentence translations**

1. 免疫現象に関する研究は、病原性微生物に対する感染防御あるいは臓器移植の際の障害の克服など臨床的な要請に迫られつつ、目覚しい進歩を遂げてきた。免疫現象は、生体が病原性の有無によってではなく、非自己の成分を認識してこれに反応することを通じて恒常性を維持する仕組みと理解してよい。この理解に立てば免疫現象が“高等な”哺乳類の専有物ではないという予想が成立する。

2. 近年の植物バイオテクノロジーの進展は目覚しいものがあり、プロトプラスト培養、細胞融合、遺伝子操作とさまざまな植物でこれらの新しい技術が研究開発されている。植物バイオテクノロジーは特に企業が将来ビジネスにつながるという見込みで、力を入れている。日本の企業は基礎研究と応用研究の両方をやっているが、開発された技術が外部に漏れないよう機密保持に神経をとがらせている。

3．動物細胞では粗面小胞体でペプチドが合成されると、小胞体やゴルジ体で大部分のペプチドはそのAsnやSer/Thr残基にN-グリコシド型やO-グリコシド型の糖鎖の付加を受ける。これらの糖蛋白質の糖鎖は、それぞれ固有な構造を介して種々の細胞間接着でそのリガンドとして、またリソゾーム酵素の細胞内器官への局在化のシグナルとして直接機能したり、蛋白質の機能制御やペプチドをプロテアーゼから保護するなど間接的にも機能している。

4．細胞や個体が平常温度より5〜10℃程度高い温度変化を急激に受けるとき、熱ショック蛋白質(HSP)の合成が誘導される。この蛋白質の合成は熱ショックのほか、さまざまな化学物質、例えば電子伝達系の阻害剤、遷移金属、SH試薬、エタノールなどによっても誘導される。そのためHSPはストレス蛋白質とも呼ばれる。

5．培養している細胞を細胞周期の一定時期にそろえることにより、個々の細胞で起きている現象を細胞集団全体の現象としてとらえることができる。このような培養法を同調培養と呼ぶ。同調培養を得る方法としては、細胞の物理的な性質を利用して特定の時期の細胞を分別する方法、DNAの前駆体合成を阻害し、DNA合成期に細胞をそろえる方法などがある。

6．微生物を含む液体を微細な孔径の沪過器に通すことにより、微生物を除去して無菌の液体を得ることができる。この方法は熱に弱い培地成分などの滅菌に用いられる。以前は石綿の平板からなるザイツ沪過器などが用いられていたが、現在ではニトロセルロースなどからなるメンブランフィルターが用いられる。その孔径には各種あり、通常のものは細菌以上の大きさのものだけを除去するが、ウイルスを除去できるものもある。

7．真核細胞の内部には多くの構造が認められるが、その中で一定の機能をもつ構造単位が細胞小器官と呼ばれている。個体がいくつかの機能をもつ器官から成り立っていることとの類似性からこの名称が生まれ、細胞内にも器官と似た機能分担が見いだされる。例えば呼吸を$O_2$の消費、$CO_2$の発生と考えれば、ミトコンドリアがこの役割を担い、消化を生体高分子の加水分解反応と考えればリソソームが消化管に相当するといえよう。

8．酸化的リン酸化においては、酸化還元反応に共役してリン酸化反応すなわちADPと$P_i$からのATP合成反応が行われるが、この時ADPあるいは$P_i$が存在しないと、呼吸速度も著しく減少する。この現象を呼吸調節という。また、このADP(あるいは$P_i$)の存在しない遅い呼吸を状態4呼吸、ADPと$P_i$の存在する時の速い呼吸を状態3呼吸と呼ぶ。

9．生体内で営まれる化学反応は驚くほど多数であるが、それらを反応の種類によって整理してみると、実はそれほど多くはない。ここでは、糖を中心とした経路について述べるが、生体物質の合成と分解は、一連の化学反応によって達成されており、ごく限られた代謝経路を中心として六つに分類することができる。その一つは解糖系である。グルコースが嫌気的に分解して、ピルビン酸、ATP、NADH各2分子を生成する経路で、主として細胞質で進行する。

10. 天然培地に使用されている天然物質として、血清、血漿、組織抽出液、蛋白質化水分解物、酵母エキス、血清分画などがある。天然培地による培養例として、ヒト胎児細胞の培養に羊水(90%)、ウマ血清(5%)、ウシ胎児抽出液(5%)を用いた成功例がある。これら天然物質は製品ロットにより活性が異なったり、天然物質の調製操作の段階で増殖活性を失わせてしまうといった問題がある。

| | | |
|---|---|---|
| 病原性 | ビョウゲンセイ | pathogenicity |
| 防御 | ボウギョ | defense |
| 臓器 | ゾウキ | organ |
| 移植 | イショク | transplantation, grafting |
| 克服 | コクフク | conquest |
| 臨床的な | リンショウテキな | clinical |
| 要請 | ヨウセイ | requirement, demand |
| 迫る | せまる | to compel, to drive |
| 目覚しい | めざましい | remarkable, striking |
| 進歩 | シンポ | progress |
| 遂げる | とげる | to accomplish, to achieve |
| 有無 | ウム | existence |
| 非自己 | ヒジコ | not of the self |
| 認識 | ニンシキ | recognition |
| 恒常性 | コウジョウセイ | homeostasis |
| 維持 | イジ | support |
| 仕組み | シくみ | mechanism, device |
| 専有物 | センユウブツ | exclusive possession |
| 予想 | ヨソウ | conjecture |
| 近年 | キンネン | recent years |
| 進展 | シンテン | progress |
| 開発 | カイハツ | development (as in R&D) |
| 企業 | キギョウ | enterprise, corporation |

| | | |
|---|---|---|
| 将来ビジネス | ショウライビジネス | future business |
| つながる | | to be connected, to be related |
| 見込み | みこみ | outlook |
| 両方 | リョウホウ | both |
| 漏れる | もれる | to be disclosed; to leak out |
| 機密保持 | キミツホジ | maintaining secrecy |
| 神経をとがらす | シンケイをとがらす | to be nervous |
| 残基 | ザンキ | residue |
| -型 | がた | type |
| 糖鎖 | トウサ | sugar chain |
| 付加 | フカ | addition |
| 細胞間接着 | サイボウカンセッチャク | intercellular adhesion |
| 局在化 | キョクザイカ | localization |
| 制御 | セイギョ | regulation, control |
| 保護 | ホゴ | protection |
| 急激に | キュウゲキに | abruptly |
| 誘導 | ユウドウ | induction |
| 電子伝達系 | デンシデンタツケイ | electron transport system |
| 遷移金属 | センイキンゾク | transition metal |
| 試薬 | シヤク | reagent |
| そろえる | | to put in order, to make uniform |
| とらえる | | to grasp, capture |
| 分別 | ブンベツ | fractionation |

| | | |
|---|---|---|
| 微細な | ビサイな | fine, minute |
| 孔径 | コウケイ | pore diameter, pore size |
| 除去 | ジョキョ | removal |
| 滅菌 | メッキン | sterilization |
| 石綿 | いしわた | asbestos |
| ザイツ | | Seitz |
| ニトロセルロース | | nitrocellulose |
| メンブランフィルター | | membrane filter |
| 真核細胞 | シンカクサイボウ | eucaryotic cell |
| 認める | みとめる | to recognize, to observe |
| 類似性 | ルイジセイ | resemblance, similarity |
| 名称 | メイショウ | name |
| 分担 | ブンタン | sharing (a burden) |
| 担う | になう | to bear (a burden) |
| 消化 | ショウカ | digestion |
| 消化管 | ショウカカン | alimentary canal |
| 共役 | キョウヤク | coupling |

| | | |
|---|---|---|
| 著しい | いちじるしい | conspicuous, striking |
| 遅い | おそい | slow |
| 営む | いとなむ | to carry on, to conduct |
| 驚く | おどろく | to be surprised |
| 整理 | セイリ | (re)arrangement, (re)ordering |
| 一連 | イチレン | a series, a chain |
| 限る | かぎる | to be limited |
| 六つ | むっつ | six |
| 解糖系 | カイトウケイ | glycolytic pathway |
| 血漿 | ケッショウ | (blood) plasma |
| 抽出液 | チュウシュツエキ | extract |
| エキス | | extract |
| 分画 | ブンカク | fraction |
| 胎児 | タイジ | fetus; embryo |
| 羊水 | ヨウスイ | amniotic fluid |
| ウマ | 【馬】 | horse |
| ウシ | 【牛】 | cow |
| 成功 | セイコウ | success |

**Translations for Ex. 3.4**

1. Agar or agarose, one of the constituents in agar, will dissolve in water when heated. If we hold these aqueous solutions at 40–50°C, add a biocatalyst and cool, an immobilized biocatalyst is obtained.
2. In aerobic metabolism 38 molecules of ATP are produced from one molecule of glucose via the citric acid cycle, but in glycolysis or in alcohol fermentation, which are [forms of] anaerobic metabolism, only two molecules of ATP are produced [from one molecule of glucose].
3. A strictly anaerobic bacterium is a bacterium that cannot grow in the presence of oxygen. It acquires energy through anaerobic fermentation of organic substrates.
4. In terms of evolution it is thought that anaerobic respiration appeared first and aerobic respiration appeared later. As a mode of acquiring energy, the efficiency of anaerobic respiration is lower than that of aerobic respiration.
5. A bacterium that can grow at temperatures of 75°C or higher is called an extreme thermophile. One that can grow between 55°C and 75°C is called a moderate thermophile.
6. Fish that live in the polar regions have been extensively researched as psychrophilic organisms. For example, it is known that fish produce a peptide that acts as a freezing point depressant to prevent the freezing of cells.
7. In his clonal selection theory for the mechanism of antibody production F.M. Burnet hypothesized the existence of forbidden clones. He considered that when autoimmunity did not function in normal animals it was because the forbidden clones had disappeared.
8. Metabolism is composed of individual, contiguous reactions. However, those reactions are catalyzed by different, individual enzymes. Thus, metabolic regulation is achieved by varying the functions of those enzymes, particularly the enzyme for the rate-limiting step.
9. It is thought that gene expression is regulated principally at the two steps of transcription from DNA to mRNA and translation from mRNA to protein. The latter step is called translational control.
10. In contrast to the virulent phage, which forms a transparent plaque on a flat plate of sensitive cells, the temperate phage forms a turbid plaque on a flat plate that has been seeded with sensitive cells.

| | | |
|---|---|---|
| 鏡 | キョウ | mirror |
| | かがみ | mirror |
| 蛍 | ケイ | firefly |
| | ほたる | firefly |
| 顕 | ケン | manifest |
| 雑 | ザツ | miscellaneous |
| 紫 | シ | purple, violet |
| | むらさき | purple, violet |
| 修 | シュウ | completion; repair |
| 飾 | ショク | decoration, ornament |
| 復 | フク | return, restoration |
| 片 | ヘン | sheet, leaf; piece |
| | かた- | single-, one-sided |
| 優 | ユウ | superior |

鏡 蛍
顕 雑
紫 修
飾 復
片 優

## 鏡

| | | |
|---|---|---|
| 位相差顕微鏡 | イソウサケンビキョウ | phase contrast microscope |
| 蛍光顕微鏡 ★ | ケイコウケンビキョウ | fluorescence microscope |
| 顕微鏡 ★ | ケンビキョウ | microscope |
| 光学顕微鏡 | コウガクケンビキョウ | light microscope |
| 電子顕微鏡 ★ | デンシケンビキョウ | electron microscope |
| 倒立顕微鏡 | トウリツケンビキョウ | inverted microscope |
| 微分干渉顕微鏡 | ビブンカンショウケンビキョウ | differential interference microscope |
| 免疫電子顕微鏡法 | メンエキデンシケンビキョウホウ | immunoelectron microscopy |

## 蛍

| | | |
|---|---|---|
| 間接蛍光抗体法 | カンセツケイコウコウタイホウ | indirect fluorescent antibody technique |
| 間接免疫蛍光法 | カンセツメンエキケイコウホウ | indirect immuno-fluorescence technique |
| 蛍光 ★ | ケイコウ | fluorescence |
| 蛍光強度 | ケイコウキョウド | fluorescence intensity |
| 蛍光顕微鏡 ★ | ケイコウケンビキョウ | fluorescence microscope |
| 蛍光抗体法 ★ | ケイコウコウタイホウ | fluorescent antibody technique |
| 蛍光光度法 | ケイコウコウドホウ | fluorography |
| 蛍光収率 | ケイコウシュウリツ | fluorescence efficiency |
| 蛍光収量 | ケイコウシュウリョウ | fluorescence yield |
| 蛍光消光 | ケイコウショウコウ | fluorescence quenching |
| 蛍光定量法 | ケイコウテイリョウホウ | fluorimetry |
| 蛍光発光 | ケイコウハッコウ | fluorescence emission |
| 蛍光標示式細胞分取器 | ケイコウヒョウジシキサイボウブンシュキ | fluorescence-activated cell sorter |
| 蛍光プローブ | ケイコウプローブ | fluorescent probe |
| 蛍光偏光解消 | ケイコウヘンコウカイショウ | fluorescence depolarization |
| 遅延蛍光 | チエンケイコウ | delayed fluorescence |
| 分光蛍光光度計 | ブンコウケイコウコウドケイ | spectrophotofluorometer |
| 蛍 | ほたる | firefly |
| 免疫蛍光法 | メンエキケイコウホウ | immunofluorescence technique |
| 流動微小蛍光測定 | リュウドウビショウケイコウソクテイ | flow microfluorometry |

## 顕

| | | |
|---|---|---|
| 位相差顕微鏡 | イソウサケンビキョウ | phase contrast microscope |
| 蛍光顕微鏡 ★ | ケイコウケンビキョウ | fluorescence microscope |
| 顕性 | ケンセイ | manifest |
| 顕微鏡 ★ | ケンビキョウ | microscope |
| 顕微操作 | ケンビソウサ | micromanipulation |
| 顕微注射 | ケンビチュウシャ | microinjection |
| 顕微分光分析 | ケンビブンコウブンセキ | microspectrophotometry |
| 光学顕微鏡 | コウガクケンビキョウ | light microscope |
| 電子顕微鏡 ★ | デンシケンビキョウ | electron microscope |
| 倒立顕微鏡 | トウリツケンビキョウ | inverted microscope |
| 微分干渉顕微鏡 | ビブンカンショウケンビキョウ | differential interference microscope |
| 免疫電子顕微鏡法 | メンエキデンシケンビキョウホウ | immunoelectron microscopy |

## 雑

| | | |
|---|---|---|
| 交雑 ★ | コウザツ | cross, crossing |
| 交雑種 | コウザツシュ | cross |
| 交雑不妊 | コウザツフニン | amixia |
| 細胞雑種形成 | サイボウザッシュケイセイ | cell hybridization |
| 細胞質雑種 | サイボウシツザッシュ | cytoplasmic hybrid |
| 細胞質雑種形成 | サイボウシツザッシュケイセイ | cybridization |
| 雑種 ★ | ザッシュ | hybrid |
| 雑種強勢 | ザッシュキョウセイ | heterosis |
| 雑種形成 | ザッシュケイセイ | hybridization |
| 雑種細胞 | ザッシュサイボウ | hybrid cell |
| 多遺伝子雑種 | タイデンシザッシュ | multihybrid |
| 複雑な ★ | フクザツな | complex, complicated |
| 戻し交雑 | もどしコウザツ | backcross |
| 戻し交雑育種法 | もどしコウザツイクシュホウ | backcross breeding |

## 紫

| | | |
|---|---|---|
| 遠紫外 ★ | エンシガイ | far ultraviolet |
| 紫外スペクトル | シガイスペクトル | ultraviolet spectrum |
| 紫外線 ★ | シガイセン | ultraviolet ray |
| 紫外線細胞測光法 | シガイセンサイボウソッコウホウ | ultraviolet cytophotometry |
| 紫外線照射 ★ | シガイセンショウシャ | ultraviolet irradiation |
| 紫外線突然変異生成 | シガイセントツゼンヘンイセイセイ | ultraviolet mutagenesis |
| 視紫紅 | シシコウ | porphyropsin |
| 紫膜 | シマク | purple membrane |

## 修

| | | |
|---|---|---|
| 誤り勝ちな修復 | あやまりがちなシュウフク | error-prone repair |
| 誤りのない修復 | あやまりのないシュウフク | error-free repair |
| 暗修復 | アンシュウフク | dark repair |
| 化学修飾 | カガクシュウショク | chemical modification |
| 修飾 ★ | シュウショク | modification |
| 修飾遺伝子 | シュウショクイデンシ | modifier |
| 修飾因子 ★ | シュウショクインシ | modifier, effector |
| 修飾塩基 | シュウショクエンキ | modified base |
| 修飾物質 | シュウショクブッシツ | modulator |
| 修復 ★ | シュウフク | repair |
| 修復酵素 | シュウフクコウソ | repair enzyme |
| 除去修復 | ジョキョシュウフク | excision repair |
| 神経修飾物質 | シンケイシュウショクブッシツ | neuromodulator |
| 複製後修復 | フクセイゴシュウフク | postreplication repair |
| 複製修復 | フクセイシュウフク | replication repair |
| 誘導的修復 | ユウドウテキシュウフク | induced repair |

## 飾

| | | |
|---|---|---|
| 化学修飾 | カガクシュウショク | chemical modification |
| 修飾 ★ | シュウショク | modification |
| 修飾遺伝子 | シュウショクイデンシ | modifier |
| 修飾因子 ★ | シュウショクインシ | modifier, effector |
| 修飾塩基 | シュウショクエンキ | modified base |
| 修飾物質 | シュウショクブッシツ | modulator |
| 神経修飾物質 | シンケイシュウショクブッシツ | neuromodulator |

## 復

| | | |
|---|---|---|
| 誤り勝ちな修復 | あやまりがちなシュウフク | error-prone repair |
| 誤りのない修復 | あやまりのないシュウフク | error-free repair |
| 暗修復 | アンシュウフク | dark repair |
| 往復機構 | オウフクキコウ | reciprocating mechanism |
| 回復 ★ | カイフク | reactivation, restoration, recovery |
| 回復熱 | カイフクネツ | recovery heat |
| 修復 ★ | シュウフク | repair |
| 修復酵素 | シュウフクコウソ | repair enzyme |
| 除去修復 | ジョキョシュウフク | excision repair |
| 反復遺伝子 | ハンプクイデンシ | repeated gene |
| 反復説 | ハンプクセツ | recapitulation theory |
| 反復配列 | ハンプクハイレツ | repetitive sequence |
| 光回復 | ひかりカイフク | photoreactivation |

| | | |
|---|---|---|
| 光回復酵素 | ひかりカイフクコウソ | photoreactivating enzyme |
| 復元 | フクゲン | renaturation |
| 復元曲線 | フクゲンキョクセン | renaturation curve |
| 複製後修復 | フクセイゴシュウフク | postreplication repair |
| 複製修復 | フクセイシュウフク | replication repair |
| 復帰 ★ | フッキ | reversion |
| 復帰突然変異 | フッキトツゼンヘンイ | reverse mutation |
| 復帰突然変異細胞 | フッキトツゼンヘンイサイボウ | reverse mutant cell |
| 復帰突然変異体 | フッキトツゼンヘンイタイ | revertant |
| 誘導的修復 | ユウドウテキシュウフク | induced repair |

# 片

| | | |
|---|---|---|
| 移植片 ★ | イショクヘン | graft, transplant |
| 移植片拒絶反応 | イショクヘンキョゼツハンノウ | graft rejection |
| 外植片 | ガイショクヘン | explant |
| 外植片培養 | ガイショクヘンバイヨウ | explant culture |
| 片側遺伝 | かたがわイデン | unilateral inheritance |
| 片方 ★ | かたホウ | one side, one of a pair |
| 砕片 | サイヘン | debris |
| 自家移植片 | ジカイショクヘン | autograft |
| 正所性移植片 | セイショセイイショクヘン | orthotopic graft |
| 挿入断片 | ソウニュウダンペン | insert |
| 組織片培養 | ソシキヘンバイヨウ | explant culture |
| 単為卵片発生 | タンイランペンハッセイ | parthenogenetic merogony |
| 断片 ★ | ダンペン | fragment |
| 断片縮合 | ダンペンシュクゴウ | fragment condensation |
| 同種移植片 | ドウシュイショクヘン | homograft |
| 雄核卵片発生 | ユウカクランペンハッセイ | andromerogony |
| 雄核卵片発生体 | ユウカクランペンハッセイタイ | andromerogen |

# 優

| | | |
|---|---|---|
| 完全優性 | カンゼンユウセイ | complete dominance |
| 半優性 | ハンユウセイ | semidominant |
| 優位 ★ | ユウイ | dominance |
| 優性 ★ | ユウセイ | dominance |
| 優性の法則 | ユウセイのホウソク | law of dominance |
| 優先種 | ユウセンシュ | dominant |
| 優良 ★ | ユウリョウ | superiority |

*SUPPLEMENTARY VOCABULARY USING KANJI FROM CHAPTER 14*

| | | |
|---|---|---|
| 一倍体 | イチバイタイ | monoploid |
| 遺伝子導入マウス | イデンシドウニュウマウス | transgenic mouse |
| 形質導入 | ケイシツドウニュウ | transduction |
| 高倍数体 | コウバイスウタイ | hyperploid |
| 固定化微生物 | コテイカビセイブツ | immobilized microorganism |
| 細胞間接着性 | サイボウカンセッチャクセイ | intercellular adhesiveness |
| 細胞接着 | サイボウセッチャク | cell adhesion |
| 三次元構造 | サンジゲンコウゾウ | three-dimensional structure |
| 三次構造 | サンジコウゾウ | tertiary structure |
| 三倍体 | サンバイタイ | triploid |
| 接合 | セツゴウ | junction, conjugation |
| 接合子 | セツゴウシ | zygote |
| 接種 | セッシュ | inoculation |
| 同時導入 | ドウジドウニュウ | cotransduction |
| 二倍性 | ニバイセイ | diploidy |
| 二倍体 | ニバイタイ | diploid |
| 微小管 | ビショウカン | microtubule |
| 微小繊維 | ビショウセンイ | microfilament |
| 微小体 | ビショウタイ | microbody |
| 微小電極 | ビショウデンキョク | microelectrode |
| 微生物 | ビセイブツ | microorganism |
| 微生物学 | ビセイブツガク | microbiology |
| 微量注射 | ビリョウチュウシャ | microinjection |
| 副経路 | フクケイロ | alternative pathway |
| 普遍形質導入 | フヘンケイシツドウニュウ | generalized transduction |
| 誘導 | ユウドウ | induction |
| 誘導酵素 | ユウドウコウソ | inducible enzyme |
| 誘導体 | ユウドウタイ | derivative |
| 誘導物質 | ユウドウブッシツ | inducer |
| 両側 | リョウがわ | bilateral |
| 類似体 | ルイジタイ | analog |

## EXERCISES

### Ex. 4.1 Matching Japanese and English terms

| | | |
|---|---|---|
| ( ) 移植片 | ( ) 紫外線照射 | ( ) 反復遺伝子 |
| ( ) 位相差顕微鏡 | ( ) 修飾遺伝子 | ( ) 片側遺伝 |
| ( ) 遠紫外 | ( ) 修飾塩基 | ( ) 免疫蛍光法 |
| ( ) 蛍光顕微鏡 | ( ) 修復酵素 | ( ) 戻し交雑 |
| ( ) 光回復酵素 | ( ) 多遺伝子雑種 | ( ) 優性 |

1. backcross
2. dominance
3. far ultraviolet
4. fluorescence microscope
5. graft, transplant
6. immunofluorescence technique
7. modified base
8. modifier
9. multihybrid
10. phase contrast microscope
11. photoreactivating enzyme
12. repair enzyme
13. repeated gene
14. ultraviolet irradiation
15. unilateral inheritance

### Ex. 4.2 KANJI with similar structural elements

Look carefully at each of the two KANJI on the left, and note which structural element is common to both. Combine each KANJI on the left with the appropriate KANJI on the right to make a meaningful JUKUGO. Each technical term that contains one or more of the 100 KANJI introduced in this book can be found in the vocabulary lists for those KANJI. Other terms can be found in one of the supplementary vocabulary lists, including Lesson 0.

| | | |
|---|---|---|
| 1. (1) 因 (2) 回 | 光( )復 | 修飾( )子 |
| 2. (1) 寒 (2) 害 | 阻( ) | ( )天 |
| 3. (1) 寒 (2) 実 | 対照( )験 | ( )冷生物 |
| 4. (1) 蛍 (2) 栄 | ( )光強度 | ( )養繁殖 |
| 5. (1) 蛍 (2) 蛋 | 間接免疫( )光法 | ( )白質生合成 |
| 6. (1) 雑 (2) 染 | 交( )種 | 常( )色体 |
| 7. (1) 紫 (2) 素 | ( )外線 | 調節酵( ) |
| 8. (1) 濁 (2) 沈 | ( )降 | ( )度 |
| 9. (1) 濁 (2) 溶 | 懸( )培養 | 濁り( )菌斑 |
| 10. (1) 脱 (2) 説 | ( )水酵素 | ゆらぎ( ) |
| 11. (1) 復 (2) 後 | 翻訳( )調節 | 誘導的修( ) |
| 12. (1) 冷 (2) 次 | ( )却遠心機 | 二( )元電気泳動 |

**Ex. 4.3 Matching Japanese technical terms with definitions**

Read each definition carefully, and then choose the appropriate technical term. Words that you have not yet encountered are listed following the definitions.

( ) 蛍光強度
( ) 蛍光顕微鏡
( ) 雑種
( ) 紫外線照射
( ) 修飾因子
( ) 修復酵素
( ) 組織片培養
( ) 電子顕微鏡
( ) 反復遺伝子
( ) 半優性

1. 電子線及び可変電磁場を利用した電磁レンズを用いて、微細な構造体の拡大像を得る機器。
2. 顕微鏡下において試料に励起光を入射し、フイルターなどにより励起光を除き、試料の発する蛍光を視野に収めるよう工夫された装置。
3. 蛍光として放射される光の強さ。
4. 遺伝的に異なった細胞または個体の交雑によって得られる細胞または個体。
5. 物質を200〜400nmの波長をもつ電磁波にさらすこと。
6. 酵素の触媒部位とは別のアロステリック部位に可逆的に結合し、酵素活性を促進または阻害するような化合物。
7. DNAの損傷を元通りに復元する反応系に働く酵素。
8. ゲノム内に何百もの単位が繰返されて存在し、しかもすべての単位がまったく同じ遺伝子であると考えられているもの。
9. 組織の断片をそのまま培養する広義の組織培養の一様式。
10. ヘテロ個体の表現型が対応する二つのホモ個体のちょうど中間の場合は、優性の割合が1/2になること。

| | | |
|---|---|---|
| 可変- | カヘン- | variable |
| 拡大像 | カクダイゾウ | magnified image |
| 試料 | シリョウ | sample, specimen |
| 励起光 | レイキコウ | excitation light |
| 入射する | ニュウシャする | to be incident |
| 視野 | シヤ | field of vision |
| 収める | おさめる | to gather |
| 工夫 | クフウ | design, scheme |
| 波長 | ハチョウ | wave length |
| 電磁波 | デンジハ | electromagnetic wave |
| 損傷 | ソンショウ | damage, injury |
| 元通り | もとどおり | as it was before |
| 何百も | なんビャクも | hundreds of ... |
| ちょうど | | exactly |

**Ex. 4.4 Sentence translations**

Read each sentence carefully, and then translate it. Words that you have not yet encountered are listed following the sentences.

1. 電子顕微鏡用の試料は高真空下の電子顕微鏡内で電子線照射を受け、電子線障害を受ける。生体試料は生きていた状態とは非常に異なった状態に作成されているので、像解釈は注意深く行うべきである。
2. 光の吸収によって励起された分子や原子が光を放出する過程のうち、発光が起こる最初の状態と発光の終った状態とが同じ多重度をもっている場合を蛍光と呼ぶ。

3. 蛍光の特性を利用し、生体物質やそれをとりまく環境の性質を調べるために、蛍光プローブという物質を用いる場合が多い。
4. ある品種に存在する優良形質を他の優良品種に導入する場合、戻し交雑育種法は有効である。家畜の改良や作物の耐寒性の導入などに利用される。
5. 細胞雑種形成は遺伝子地図の作成や、ヒト遺伝病の発症機構の解明などに広く利用されるようになった。
6. 光学顕微鏡により、薄片試料中の物質の微量分析を行うのが顕微分光分析であるが、特に紫外線を光源として、細胞中の物質分布を測定するのは紫外線細胞測光法と呼ばれている。
7. 修飾塩基は天然の核酸中に通常の塩基以外に微量に含まれる塩基誘導体であり、その多くは核酸の複製や転写後に酵素的に修飾されて生成する。
8. 一般に電離性放射線やアルキル化剤による損傷の修復時にはDNA鎖上の短い部分が切除・再合成されるが、紫外線損傷の場合は長い領域にわたって修復が行われる。
9. ある個体の一部を同一個体または異なる個体に移植する場合、提供される組織または臓器を移植片と呼ぶ。移植片に含まれる組織適合性抗原のことを移植抗原と呼び、受容者のそれとの差異、特に主要組織適合性抗原系における差異の程度に応じて受容者の免疫反応が起こる。
10. 優性や劣性度は、遺伝子それ自体の性質ではなく、特定の遺伝子型の全体的な反応系のうちで問題とする遺伝子座の作用の結果とみるべきである。

| | | |
|---|---|---|
| 真空 | シンクウ | vacuum |
| 非常に | ヒジョウに | extremely |
| 作成 | サクセイ | drawing up, preparing |
| 像解釈 | ゾウカイシャク | image interpretation |
| 注意深く | チュウイぶかく | with great caution |
| 吸収 | キュウシュウ | absorption |
| 励起 | レイキ | excitation |
| 終る | おわる | to end |
| 多重度 | タジュウド | multiplicity |
| とりまく | | to surround |
| 品種 | ヒンシュ | breed, variety |
| 有効 | ユウコウ | effective |
| 家畜 | カチク | domestic animals |
| 改良 | カイリョウ | improvement |
| 地図 | チズ | map |
| 遺伝病 | イデンビョウ | hereditary disease |
| 発症 | ハッショウ | onset of a disease |
| 薄片 | ハクヘン | lamina, thin section |
| 光源 | コウゲン | light source |
| 複製 | フクセイ | replication, duplication |
| アルキル化剤 | アルキルカザイ | alkylating agent |
| DNA 鎖 | DNA サ | chain |
| 短い | みじかい | short |
| 切除 | セツジョ | excision |
| -にわたって | | extending over |
| 提供 | テイキョウ | offer, supply |
| 組織適合性 | ソシキテキゴウセイ | histocompatibility |
| 移植抗原 | イショクコウゲン | transplantation antigen |
| 受容者 | ジュヨウシャ | recipient |
| 劣性度 | レッセイド | recessivity |
| 遺伝子型 | イデンシがた | genotype |
| 問題 | モンダイ | problem, issue |
| 座 | ザ | locus |

#### Ex. 4.5 Additional dictionary entries

| | | |
|---|---|---|
| ( ) 遺伝子導入マウス | ( ) 接合 | ( ) 微小管 |
| ( ) 形質導入 | ( ) 接種 | ( ) 微生物学 |
| ( ) 固定化微生物 | ( ) 二倍体細胞 | ( ) 誘導酵素 |
| ( ) 細胞接着 | | |

1. 新しい動物体や培地に細菌、細胞あるいはウイルス、ワクチンなどを植え込むこと。
2. 構成酵素と違い、特定の基質の存在に応じて生体中の酵素分子の合成速度が変化する酵素。
3. 酵素を工業的に触媒として利用するための、酵素を含む菌体や複雑な酵素系をもつ微生物または微生物細胞の固定化。
4. 受精直後のマウス卵に特定の遺伝子を注入した後マウスの子宮に移植するという方法で誕生したマウスで、体内で注入した遺伝子が機能しているもの。
5. 真核細胞に広く存在する、いろいろな型の細胞内運動に関与している外径25nmの管状構造。
6. 生体を構成する体細胞で、両親から由来する相同染色体を一対ずつもった細胞。
7. 多細胞生物を構成する細胞相互間の糖蛋白質、カルシウムブリッジ、またはファンデルワールス力によるくっつき。
8. 単細胞生物における、交配型が逆の同一種の生物が対合し、遺伝物質を交換する有性生殖過程。
9. 微生物に関する分野で、微生物を研究する手法によって、微生物遺伝学、微生物生化学、免疫学などの領域のある学問。
10. ファージを介して遺伝形質が供与菌から受容菌に受け渡されること。

| | | | | | |
|---|---|---|---|---|---|
| 植え込む | うえこむ | to plant, to introduce | ブリッジ | | bridge |
| 工業 | コウギョウ | industry | くっつく | | to stick to, to adhere to |
| 子宮 | シキュウ | womb, uterus | 交配型 | コウハイがた | mating type |
| 誕生 | タンジョウ | birth | 逆の | ギャクの | reverse, contrary |
| 外径 | ガイケイ | outer diameter | 受け渡す | うけわたす | to transfer, to deliver |
| 両親 | リョウシン | parents | | | |

**Translations for Ex. 4.4**

1. The sample used in an electron microscope is irradiated with an electron beam under high vacuum within the electron microscope, and is damaged by the electon beam. Because a biological sample is prepared in a state that is very different from the living state, interpretation of the image must be carried out with great caution.
2. In the process by which a molecule or an atom that has been excited by the absorption of light radiates light, fluorescence describes the situation when the initial state at which luminescence occurs has the same multiplicity as the state after luminescence has ended.
3. A substance known as a fluorescent probe is often used in order to make use of the characterisitcs of fluorescence to investigate a biological substance or the properties of the environment that surrounds it.
4. When a superior trait that exists in a certain breed is introduced into another superior breed the method of backcross breeding is effective. It is employed for the improvement of domestic animals, for the introduction of cryotolerance into crops, and so forth.
5. Cell hybridization has come to be widely employed for drawing up gene maps, for elucidating the onset mechanism of hereditary diseases in humans, and so forth.
6. Carrying out the microanalysis of a substance in a laminar sample by means of an optical microscope is microspectrophotometry. In particular, measuring the distribution of materials within a cell using ultraviolet light as a light source is called ultraviolet cytophotometry.
7. A modified base is a base derivative that is found in trace quantities in a natural nucleic acid in addition to the ordinary bases. Most modified bases are produced through enzymatic modification following replication or transcription of the nucleic acid.
8. Generally, during the repair of damage caused by ionizing radiation or an alkylating agent short portions of the DNA chain are excised or resynthesized. However, in the case of ultraviolet damage repairs take place over long regions.
9. When one portion of an individual is grafted onto the same individual or a different individual, the tissue or organ that we supply is called the graft. The histocompatibility antigens contained in the graft are called transplantation antigens. An immune reaction by the recipient occurs in accordance with the difference between these antigens and those of the recipient, particularly for the major histocompatibility system
10. Dominance and recessivity are not properties of a gene itself. They must be viewed as the results of the action of the gene locus at issue within the entire reaction system of a particular genotype.

| | | |
|---|---|---|
| 株 | かぶ | strain (of a species), stock |
| 感 | カン | sensing, feeling |
| | カン(じる) | to feel, sense {v.t.} |
| | カン(じ) | feeling, sensation |
| 凝 | ギョウ | coagulation |
| 枝 | シ | branch |
| | えだ | branch |
| 旋 | セン | turning, rotating |
| 促 | ソク | stimulating, promoting |
| 毒 | ドク | poison, toxin |
| 皮 | ヒ | skin, hide |
| | かわ | skin, hide |
| 病 | ビョウ | disease |
| 母 | ボ | mother |
| | はは | mother |

株 感

凝 枝

旋 促

毒 皮

病 母

# 株

| | | |
|---|---|---|
| 亜株 | アかぶ | substrain |
| 温度感受性変異株 | オンドカンジュセイヘンイかぶ | temperature sensitive mutant |
| 菌株 ★ | キンかぶ | strain |
| 原株 | ゲンかぶ | original strain |
| 高温感受性変異株 | コウオンカンジュセイヘンイかぶ | temperature sensitive mutant |
| 細胞株 | サイボウかぶ | cell strain |
| 低温感受性変異株 | テイオンカンジュセイヘンイかぶ | cold sensitive mutant |
| 変異株 ★ | ヘンイかぶ | mutant |
| 野生株 ★ | ヤセイかぶ | wild strain |

# 感

| | | |
|---|---|---|
| 追いうち感染 | おいうちカンセン | superinfection |
| 温度感受性変異株 | オンドカンジュセイヘンイかぶ | temperature sensitive mutant |
| 化学感覚毛 | カガクカンカクモウ | chemosensory hair |
| 感覚 ★ | カンカク | sense, sensation |
| 感覚器官 | カンカクキカン | sense organ, sensory organ |
| 感覚性ニューロン | カンカクセイニューロン | sensory neuron |
| 感作 ★ | カンサ | sensitization |
| 感作リンパ球 | カンサリンパキュウ | sensitized lymphocyte |
| 感受性 | カンジュセイ | sensitive, sensitivity |
| 感染 ★ | カンセン | infection |
| 感染性核酸 | カンセンセイカクサン | infectious nucleic acid |
| 感染多重度 | カンセンタジュウド | multiplicity of infection |
| 感染病原菌 | カンセンビョウゲンキン | infectious or pathogenic microbe |
| 高温感受性変異株 | コウオンカンジュセイヘンイかぶ | temperature sensitive mutant |
| 光増感 | コウゾウカン | photosensitization |
| 視感度曲線 | シカンドキョクセン | visibility curve |
| 受動感作 | ジュドウカンサ | passive sensitization |
| 重感染 | ジュウカンセン | superinfection |
| 遷延感作 | センエンカンサ | prolonged sensitization |
| 増感 | ゾウカン | sensitization |
| 他感作用 | タカンサヨウ | allelopathy |
| 多重感染 | タジュウカンセン | multiple infection |
| 脱感作 | ダツカンサ | desensitization |

| | | |
|---|---|---|
| 単感染 | タンカンセン | single infection |
| 通性感染 | ツウセイカンセン | facultative infection |
| 低温感受性変異株 | テイオンカンジュセイヘンイかぶ | cold sensitive mutant |
| 半感染量 | ハンカンセンリョウ | fifty percent infection dose |
| 非感受性 | ヒカンジュセイ | insensitive, insensitivity |
| 不感時間 | フカンジカン | dead time |
| 不稔感染 | フネンカンセン | abortive infection |
| 放射線感受性 | ホウシャセンカンジュセイ | radiosensitivity |
| 溶菌感染 | ヨウキンカンセン | lytic infection |

## 凝

| | | |
|---|---|---|
| 間接凝集反応 | カンセツギョウシュウハンノウ | indirect agglutination |
| 間接血球凝集 | カンセツケッキュウギョウシュウ | indirect hemagglutination |
| 寒冷凝集素 | カンレイギョウシュウソ | cold agglutinin |
| 寒冷赤血球凝集素 | カンレイセッケッキュウ ギョウシュウソ | cold hemagglutinin |
| 凝塊 | ギョウカイ | clump, clot |
| 凝結 ★ | ギョウケツ | coagulation |
| 凝固 ★ | ギョウコ | coagulation, solidification |
| 凝固血漿 | ギョウコケッショウ | plasma clot |
| 凝固血漿培養 | ギョウコケッショウバイヨウ | plasma clot culture |
| 凝固点降下 | ギョウコテンコウカ | freezing point depression |
| 凝集 ★ | ギョウシュウ | agglutination, aggregation |
| 凝集因子 | ギョウシュウインシ | aggregating factor |
| 凝集塊 | ギョウシュウカイ | aggregate |
| 凝集原 | ギョウシュウゲン | agglutinogen |
| 凝集素 | ギョウシュウソ | agglutinin |
| 凝集体 | ギョウシュウタイ | aggregate |
| 凝析 | ギョウセキ | coagulation |
| 凝乳酵素 | ギョウニュウコウソ | milk-clotting enzyme |
| 血液凝固 | ケツエキギョウコ | blood coagulation, blood clotting |
| 血液凝固因子 | ケツエキギョウコインシ | blood coagulation factor, blood clotting factor |
| 血球凝集 | ケッキュウギョウシュウ | hemagglutination |
| 血球凝集素 | ケッキュウギョウシュウソ | hemagglutinin |
| 血球凝集阻止 | ケッキュウギョウシュウソシ | hemagglutination inhibition |
| 細胞凝集素 | サイボウギョウシュウソ | cytoagglutinin |
| 受動凝集 | ジュドウギョウシュウ | passive agglutination |
| 受動血球凝集 | ジュドウケッキュウギョウシュウ | passive hemagglutination |
| 植物凝集素 | ショクブツギョウシュウソ | phytohemagglutinin |
| 多凝集反応 | タギョウシュウハンノウ | polyagglutination |

| | | |
|---|---|---|
| 直接凝集反応 | チョクセツギョウシュウハンノウ | direct agglutination |
| 同種凝集素 | ドウシュギョウシュウソ | isoagglutinin |
| 同種血球凝集 | ドウシュケッキュウギョウシュウ | isohemagglutination |
| 乳凝 | ニュウギョウ | curdling |

## 枝

| | | |
|---|---|---|
| 枝切り酵素 | えだきりコウソ | debranching enzyme |
| 枝つくり酵素 ★ | えだつくりコウソ | branching enzyme |
| 枝分かれ ★ | えだわかれ | branching |
| 脱分枝酵素 | ダツブンシコウソ | debranching enzyme |
| 分枝アミノ酸 | ブンシアミノサン | branched chain amino acid |
| 分枝系 | ブンシケイ | clone |
| 分枝酵素 ★ | ブンシコウソ | branching enzyme |
| 分枝点移動 | ブンシテンイドウ | branch migration |

## 旋

| | | |
|---|---|---|
| 右旋性 | ウセンセイ | dextrorotatory, dextrorotation |
| 左旋性 | サセンセイ | levorotatory, levorotation |
| 残基旋光度 | ザンキセンコウド | residue rotation |
| 旋回培養 ★ | センカイバイヨウ | gyratory culture |
| 旋光 ★ | センコウ | optical rotation |
| 旋光強度 | センコウキョウド | rotatory strength |
| 旋光計 | センコウケイ | polarimeter |
| 旋光度 | センコウド | optical rotation |
| 旋光分析 | センコウブンセキ | polarimetry |
| 二重螺旋 (or 二重らせん) | ニジュウラセン | double helix |
| 比旋光度 | ヒセンコウド | specific rotation |
| 分光旋光計 | ブンコウセンコウケイ | spectropolarimeter |
| 分子旋光度 | ブンシセンコウド | molecular rotation |
| 変旋光 | ヘンセンコウ | mutarotation |
| 螺旋 (or らせん) ★ | ラセン | helix |
| 螺旋含量 (or らせん含量) | ラセンガンリョウ | helix content |
| 螺旋糸 (or らせん糸) | ラセンシ | spireme |
| 螺旋軸 (or らせん軸) | ラセンジク | screw axis |

## 促

| | | |
|---|---|---|
| 造血促進因子 | ゾウケツソクシンインシ | erythropoietic-stimulating factor |
| 促進 ★ | ソクシン | promotion, facilitation |
| 促進因子 ★ | ソクシンインシ | accelerator, promoter |

| | | |
|---|---|---|
| 促進グロブリン | ソクシングロブリン | accelerator globulin |
| 促進体 | ソクシンタイ | accelerator |
| 脱皮促進ホルモン | ダッピソクシンホルモン | molt-accelerating hormone |
| 分裂促進剤 | ブンレツソクシンザイ | mitogen |
| 免疫促進 | メンエキソクシン | immunological enhancement |

# 毒

| | | |
|---|---|---|
| 外毒素 | ガイドクソ | exotoxin |
| 蜘蛛毒 | くもドク | spider toxin |
| 解毒 | ゲドク | detoxication |
| 細胞毒 | サイボウドク | cytotoxin |
| 細胞毒性 | サイボウドクセイ | cytotoxicity |
| 蠍毒 | さそりドク | scorpion venom |
| 弱毒ウイルス | ジャクドクウイルス | attenuated virus |
| 弱毒化 | ジャクドクカ | attenuation |
| 弱毒ワクチン | ジャクドクワクチン | attenuated vaccine, inactivated vaccine |
| 消毒薬 | ショウドクヤク | disinfectant |
| 神経毒 | シンケイドク | neurotoxin |
| 心臓毒 | シンゾウドク | heart poison |
| 選択毒性 | センタクドクセイ | selective toxicity |
| 中毒 | チュウドク | addiction, intoxication |
| 腸管毒 | チョウカンドク | enterotoxin |
| 毒液 | ドクエキ | venom |
| 毒性 ★ | ドクセイ | toxic(ity) |
| 毒性蛋白質 | ドクセイタンパクシツ | toxic protein |
| 毒性ファージ | ドクセイファージ | virulent phage |
| 毒素 ★ | ドクソ | toxin |
| 毒素学 | ドクソガク | toxinology |
| 毒物 ★ | ドクブツ | poison |
| 毒物学 | ドクブツガク | toxicology |
| 内毒素 | ナイドクソ | endotoxin |
| 麦角中毒 | バクカクチュウドク | ergotism |
| 破傷風毒素 | ハショウフウドクソ | tetanus toxin |
| 蜂毒 | はちドク | bee toxin |
| 百日咳毒素 | ヒャクニチゼキドクソ | pertussis toxin |
| 河豚毒 | フグドク | fugu poison |
| 蛇毒 | ヘビドク | snake venom |
| ボツリヌス毒素 | ボツリヌスドクソ | botulinum toxin |
| 免疫毒素 | メンエキドクソ | immunotoxin |
| 有機水銀中毒 | ユウキスイギンチュウドク | organic mercury poisoning |

# 皮

| | | |
|---|---|---|
| 外皮 | ガイヒ | pellicle, tegument |
| 角皮 | カクヒ | cuticle |
| 桂皮酸 | ケイヒサン | cinnamic acid |
| 原皮質 | ゲンヒシツ | archicortex |
| 古皮質 | コヒシツ | paleocortex |
| 上皮 ★ | ジョウヒ | epithelium |
| 上皮細胞 | ジョウヒサイボウ | epithelial cell |
| 上皮性細胞 | ジョウヒセイサイボウ | epithelial-like cell |
| 上皮増殖因子 | ジョウヒゾウショクインシ | epidermal growth factor |
| 上皮組織 | ジョウヒソシキ | epithelial tissue |
| 新皮質 | シンヒシツ | neocortex |
| 腎皮質 | ジンヒシツ | renal cortex |
| 生皮 | セイヒ | rawhide |
| 体腔上皮 | タイクウジョウヒ | mesothelium |
| 大脳新皮質 | ダイノウシンヒシツ | cerebral neocortex |
| 大脳皮質 | ダイノウヒシツ | cerebral cortex |
| 脱皮 | ダッピ | ecdysis |
| 脱皮促進ホルモン | ダッピソクシンホルモン | molt-accelerating hormone |
| 脱皮ホルモン | ダッピホルモン | molting hormone |
| 中皮 | チュウヒ | mesothelium |
| 腸上皮細胞 | チョウジョウヒサイボウ | intestinal epithelium cell |
| 同皮質 | ドウヒシツ | isocortex |
| 内皮 | ナイヒ | endothelium |
| 内皮細胞 | ナイヒサイボウ | endothelial cell |
| 尿細管上皮変性 | ニョウサイカンジョウヒヘンセイ | nephrosis |
| 薄皮 | ハクヒ | pellicle |
| 皮下注射 | ヒカチュウシャ | subcutaneous injection |
| 皮質 | ヒシツ | cortex |
| 皮膚 (or 皮フ) ★ | ヒフ | cutis, skin |
| 皮膚反応 | ヒフハンノウ | skin reaction |
| 皮膚反応因子 | ヒフハンノウインシ | skin reactive factor |
| 表皮 ★ | ヒョウヒ | epidermis |
| 表皮角化細胞 | ヒョウヒカクカサイボウ | epidermal keratinocyte |
| 表皮ケラチン細胞 | ヒョウヒケラチンサイボウ | epidermal keratinocyte |
| 表皮[性] | ヒョウヒ[セイ] | epidermal |
| 副腎皮質 | フクジンヒシツ | adrenal cortex |
| 副腎皮質ホルモン | フクジンヒシツホルモン | adrenal cortical hormone |
| 不等皮質 | フトウヒシツ | allocortex |
| 辺縁皮質 | ヘンエンヒシツ | limbic cortex |
| 包皮 | ホウヒ | pellicle |

# 病

| | | |
|---|---|---|
| 一次性糖尿病 | イチジセイトウニョウビョウ | primary diabetes |
| 遺伝病 ★ | イデンビョウ | genetic disease, hereditary disease |
| 壊血病 | カイケツビョウ | scurvy |
| 感染病原菌 | カンセンビョウゲンキン | infectious or pathogenic microbe |
| 狂犬病ウイルス | キョウケンビョウウイルス | rabies virus |
| 血管性血有病 | ケッカンセイケツユウビョウ | vascular hemophilia |
| 血清病 | ケッセイビョウ | serum sickness |
| 血友病 | ケツユウビョウ | hemophelia |
| 膠原病 | コウゲンビョウ | collagen disease |
| 黒舌病 | コクゼツビョウ | black tongue |
| 心臓病患者 | シンゾウビョウカンジャ | cardiac |
| 成人T細胞白血病 | セイジンTサイボウハッケツビョウ | adult T cell leukemia |
| 精神分裂病 | セイシンブンレツビョウ | schizophrenia |
| 赤白血病細胞 | セキハッケツビョウサイボウ | erythroleukemia cell |
| 繊溶性紫斑病 | センヨウセイシハンビョウ | purpura fibrinolytica |
| 躁鬱病 | ソウウツビョウ | manic-depressive illness |
| 垂井病 | たるいビョウ | Tarui's disease |
| 糖原病 | トウゲンビョウ | glycogen storage disease |
| 糖尿病 | トウニョウビョウ | diabetes mellitus |
| 鳥白血病ウイルス | とりハッケツビョウウイルス | avian leukemia virus |
| 橋本病 | はしもとビョウ | Hashimoto's disease |
| 白血病 | ハッケツビョウ | leukemia |
| 白血病細胞 | ハッケツビョウサイボウ | leukemic cell |
| 白血病裂孔 | ハッケツビョウレッコウ | hiatus leukemicus |
| 病因 | ビョウイン | pathogenesis, cause of a disease |
| 病期 | ビョウキ | stage (of a disease) |
| 病気 ★ | ビョウキ | disease |
| 病原性 | ビョウゲンセイ | pathogenicity |
| 病原体 ★ | ビョウゲンタイ | pathogen |
| 病巣 | ビョウソウ | focus |
| 病斑 | ビョウハン | necrotic lesion |
| 浮上β病 | フジョウβビョウ | floating beta disease |
| 分子病 | ブンシビョウ | molecular disease |
| 水俣病 | みなまたビョウ | Minamata disease |
| 免疫グロブリン病 | メンエキグロブリンビョウ | immunoglobulinopathy |
| 免疫病 | メンエキビョウ | immune disease |

# 母

| | | |
|---|---|---|
| 水母 | くらげ | jellyfish |
| 酵母 ★ | コウボ | yeast |
| 酵母接合型 | コウボセツゴウがた | yeast mating type |
| 精母細胞 ★ | セイボサイボウ | spermatocyte |
| 分母 | ブンボ | denominator |
| 母液 | ボエキ | mother liquor |
| 母細胞 | ボサイボウ | metrocyte |
| 母児免疫移行 | ボジメンエキイコウ | maternal transmission of immunity |
| 母性mRNA | ボセイmRNA | maternal mRNA |
| 卵母細胞 ★ | ランボサイボウ | oocyte |

*SUPPLEMENTARY VOCABULARY USING KANJI FROM CHAPTER 15*

| | | |
|---|---|---|
| 異質染色質 | イシツセンショクシツ | heterochromatin |
| 異性化酵素 | イセイカコウソ | isomerase |
| 遺伝分析 | イデンブンセキ | genetic analysis |
| 塩基配列決定 | エンキハイレツケッテイ | sequencing |
| 塩析 | エンセキ | salting out |
| 銀染色法 | ギンセンショクホウ | silver staining |
| 交配 | コウハイ | mating, crossing |
| 勾配 | コウバイ | gradient |
| 再配列 | サイハイレツ | rearrangement |
| 細胞集団 | サイボウシュウダン | cell population |
| 細胞集落 | サイボウシュウラク | colony |
| 脂質 | シシツ | lipid |
| 集合 | ショウゴウ | aggregation, assembly |
| 集団遺伝学 | シュウダンイデンガク | population genetics |
| 集落 | シュウラク | colony |
| 生物集団 | セイブツシュウダン | biological population |
| 挿入配列 | ソウニュウハイレツ | insertion sequence |
| 挿入変異 | ソウニュウヘンイ | insertion mutation |
| 多重遺伝子族 | タジュウイデンシゾク | multigene family |
| 窒素 | チッソ | nitrogen |
| 窒素固定 | チッソコテイ | nitrogen fixation |
| 窒素固定菌 | チッソコテイキン | nitrogen-fixing bacteria |
| 窒素固定酵素 | チッソコテイコウソ | nitrogen-fixing enzyme |
| 定性分析 | テイセイブンセキ | qualitative analysis |
| 定量分析 | テイリョウブンセキ | quantitative analysis |
| 電気透析 | デンキトウセキ | electrodialysis |

| | | |
|---|---|---|
| 点変異 | テンヘンイ | point mutation |
| 透析 | トウセキ | dialysis |
| 同族体 | ドウゾクタイ | homolog |
| 任意交配 | ニンイコウハイ | random mating |
| 配位結合 | ハイイケツゴウ | coordinate bond |
| 配位子 | ハイイシ | ligand |
| 配偶子 | ハイグウシ | gamete |
| 配偶子合体 | ハイグウシゴウタイ | fertilization |
| 配列 | ハイレツ | sequence |
| 配列決定 | ハイレツケッテイ | sequencing |
| 配列相同性 | ハイレツソウドウセイ | sequence homology |
| 分子進化 | ブンシシンカ | molecular evolution |
| 分析 | ブンセキ | analysis, analytical |
| 変異 | ヘンイ | variation |
| 変異形 | ヘンイケイ | variant |
| 変異細胞 | ヘンイサイボウ | variant cell |
| 変異体 | ヘンイタイ | variant, mutant |
| メンデル集団 | メンデルシュウダン | Mendelian population |
| 立体異性 | リッタイイセイ | stereoisomerism |
| 硫酸 | リュウサン | sulfuric acid |
| リン脂質 | リンシシツ | phospholipid |
| 漏出変異体 | ロウシュツヘンイタイ | leaky mutant |

## EXERCISES

### Ex. 5.1 Matching Japanese and English terms

( ) 感染性核酸
( ) 凝集因子
( ) 血液凝固因子
( ) 光増感
( ) 酵母
( ) 細胞株

( ) 上皮細胞
( ) 神経毒
( ) 精神分裂病
( ) 旋回培養
( ) 旋光度
( ) 促進因子

( ) 脱分枝酵素
( ) 白血病
( ) 皮下注射
( ) 分枝酵素
( ) 免疫毒素

1. accelerator, promoter
2. aggregating factor
3. blood coagulation factor
4. branching enzyme
5. cell strain
6. debranching enzyme
7. epithelial cell
8. gyratory culture
9. immunotoxin
10. infectious nucleic acid
11. leukemia
12. neurotoxin
13. optical rotation
14. photosensitization
15. schizophrenia
16. subcutaneous injection
17. yeast

### Ex. 5.2 KANJI with the same ON reading

Look carefully at each of the two KANJI on the left, and note the ON reading that is common to both. Combine each KANJI on the left with the appropriate KANJI on the right to make a meaningful JUKUGO. Each technical term that contains one or more of the 100 KANJI introduced in this book can be found in the vocabulary lists for those KANJI. Other terms can be found in one of the supplementary vocabulary lists, including Lesson 0.

| | (1) | (2) | | |
|---|---|---|---|---|
| 1. | (1) 鏡 | (2) 強 | 旋光( )度 | 電子顕微( ) |
| 2. | (1) 顕 | (2) 嫌 | ( )気性生物 | ( )微注射 |
| 3. | (1) 枝 | (2) 止 | 休( )状態 | 分( )アミノ酸 |
| 4. | (1) 紫 | (2) 止 | ( )外スペクトル | 禁( )クローン |
| 5. | (1) 修 | (2) 周 | 暗( )復 | 細胞( )期 |
| 6. | (1) 飾 | (2) 殖 | 修( )物質 | 増( )培地 |
| 7. | (1) 旋 | (2) 染 | 左( )性 | 多重感( ) |
| 8. | (1) 促 | (2) 速 | 代謝( )度 | 免疫( )進 |
| 9. | (1) 培 | (2) 倍 | 斜面( )養 | 二( )体 |
| 10. | (1) 皮 | (2) 比 | 上( )増殖因子 | ( )旋光度 |
| 11. | (1) 片 | (2) 変 | ( )異株 | 外植( ) |
| 12. | (1) 優 | (2) 有 | ( )機水銀中毒 | ( )性の法則 |

**Ex. 5.3 Matching Japanese technical terms with definitions**

Read each definition carefully, and then choose the appropriate technical term. Words that you have not yet encountered are listed following the definitions.

( ) 感作
( ) 凝集
( ) 酵母
( ) 脱分枝酵素
( ) 毒素
( ) 病原体
( ) 表皮
( ) 分裂促進剤
( ) 変旋光
( ) 野生株

1. 自然界で最も高頻度に出現する表現型をもつ生物。
2. ある物質に対して、動物または細胞の反応性を特異的に高めるために、その物質を適当な方法で投与すること。
3. 液体中に分散していた細胞や細菌のような粒状体が集合して塊をつくる現象。
4. α1→6結合による分枝をもつα-1,4-グルカン中のα1→6結合を分解して直鎖グルカンを生ずる酵素の総称。
5. ある種の光学活性物質を溶媒に溶かした時、その比旋光度が経時的に変化し、やがて一定値になる現象。
6. 免疫学的には抗原非特異的にリンパ球の分裂を誘起させる物質。
7. 物理的あるいは化学的反応により生体の恒常性や生理機能に変調をきたすような物質。
8. 動物の皮膚の上皮。
9. 寄生して直接病気の原因となる生物。
10. 生活環の大部分が単細胞であり、主として出芽によって増殖する真菌類の総称。

| | | |
|---|---|---|
| 自然界 | シゼンカイ | natural world |
| 頻度 | ヒンド | frequency |
| 適当な | テキトウな | appropriate |
| 投与する | トウヨする | to administer |
| 分散 | ブンサン | dispersion |
| 塊 | かたまり | clot, lump |
| 直鎖 | チョクサ | staight chain |
| 経時的に | ケイジテキに | with the passage of time |
| やがて | | before long |
| 誘起する | ユウキする | to cause |
| 生理機能 | セイリキノウ | physiological function |
| 変調 | ヘンチョウ | irregularity, abnormality |
| きたす | | to bring about |
| 生活環 | セイカツカン | life cycle |
| 出芽 | シュツガ | budding |
| 真菌類 | シンキンルイ | *Eumycetes* |

**Ex. 5.4 Sentence translations**

Read each sentence carefully, and then translate it. Words that you have not yet encountered are listed following the sentences.

1. 細胞株とは初代培養からでもまたは細胞系からでも、選択あるいはクローニングによって特異な性質あるいは(遺伝的)標識をもつようになった培養系統を指す。
2. ファージによる感染に際して、細菌細胞1個当たりに加えたファージ粒子の数を感染多重度といい、これがmの時にr個のファージの感染を受けた菌の割合$P_r$は、

$P_r=(m^r)(e^{-m})/(r!)$で与えられる。

3. クエン酸やヘパリンなどで凝固阻止した血液から、遠心分離によって血球を除去して得た血漿と凝固因子を含む液(簡単には組織抽出液など)を混合することによって凝固血漿を得ることができる。
4. 8の字形DNA分子での分枝点の移動は長いヘテロ二本鎖部分の形成のために必須であることから、分枝点の移動は普遍的組換えにおいて不可欠な過程であると考えられている。ホリデイモデルによる遺伝子変換の説明の時にも、分枝点移動は組換え中間体の形の必須な条件であるとされている。
5. 旋回培養は、諸条件を一定にしておくと、きわめて再現性の高い、一定の大きさと形をもった細胞塊が形成されるので、細胞集合や細胞選別の研究に広く用いられている。
6. 細菌の生産する毒素のうち、外毒素は高分子蛋白質で熱に不安定であり、毒性はきわめて強い。古典的なジフテリア毒素、破傷風毒素およびボツリヌス毒素などがその例である。
7. 化学療法剤の場合には、人間または家畜のような有益種には害を与えず、寄生虫や感染病原菌のような有害種にのみ選択毒性を発揮するのが望ましい。
8. 上皮増殖因子(EGF)は種々の上皮組織に作用してその成長を特異的に促進する化学物質である。神経成長因子(NGF)と類似して唾液腺より分泌されるが、このEGFはインスリン受容体または神経成長因子受容体には結合しない。
9. あらゆる病気は、遺伝要因と環境要因の複雑な絡み合いから生ずると考えられているが、そのうち特に遺伝要因が病気の発現と直結しているものを遺伝病という。
10. 多くの動物では排卵まで卵母細胞は巨大な核をもつ第一成熟分裂前期の状態でとどまっており、この前期はきわめて長い過程で、この間mRNAの合成は盛んである。

| | | |
|---|---|---|
| 初代 | ショダイ | first generation |
| 細胞系 | サイボウケイ | cell line |
| 系統 | ケイトウ | line, stock |
| 際して | サイして | when |
| 簡単 | カンタン | simplicity |
| 混合 | コンゴウ | mixture |
| 8の字形 | ハチのジがた | figure 8 |
| ヘテロ二本鎖 | ヘテロニホンサ | heteroduplex |
| 普遍的 | フヘンテキ | general, universal |
| 不可欠な | フカケツな | indispensable |
| 変換 | ヘンカン | conversion |
| 諸条件- | ショジョウケン | various conditions |
| 再現性 | サイゲンセイ | reproducibility |
| 細胞塊 | サイボウカイ | cell aggregate |
| 選別 | センベツ | sorting |
| 古典的な | コテンテキな | classical |
| 化学療法剤 | カガクリョウホウザイ | chemotherapeutic agent |
| 有益種 | ユウエキシュ | useful species |
| 寄生虫 | キセイチュウ | parasite |
| 発揮 | ハッキ | manifestation |
| 望ましい | のぞましい | to be desirable |
| 唾液腺 | ダエキセン | salivary gland |
| 分泌 | ブンピツ | secretion |
| 絡み合い | からみあい | entanglement |
| 排卵 | ハイラン | ovulation |
| 巨大な | キョダイな | huge |
| 成熟 | セイジュク | maturation |
| とどまる | | to halt |
| 盛ん | さかん | flourishing, active |

**Ex. 5.5 Additional dictionary entries**

( ) 異性化酵素
( ) 遺伝分析
( ) 交配
( ) 挿入配列
( ) 多重遺伝子族
( ) 点変異
( ) 配列相同性
( ) 変異細胞
( ) リン脂質
( ) 漏出変異体

1. ある遺伝形質に関与する遺伝子の数、染色体上の位置、表現型に及ぼす影響などの決定。
2. ある表現形質に関して、元の細胞と異なる細胞集団で、一般的にその形質が不安定なので、環境条件によって元に戻ることが多いもの。
3. 元来は染色体上で強度に連鎖し、機能的に関連のある反復性をもったDNAの配列または遺伝子群。
4. 生物種間、または種内の二つ以上の蛋白質や核酸の一次構造を比較する際、配列の類似性が偶然の類似性より高い場合に、これらのアミノ酸配列または塩基配列が示す性質。
5. 接合体を形成するため、2個体間で受粉あるいは受精を行うこと。
6. 染色体、プラスミド、ファージDNAなど種々のレプリコンの間を移動する移動性DNAの一種で、移動性DNAの中では最も小さく簡単な構造をもっているもの。
7. DNA上の1塩基対(単鎖生物では1塩基)が異なる塩基対に起き換わった変異。
8. 突然変異を起こした遺伝子産物が完全に失活していないで、ある程度野生型の機能を保持している変異体。
9. 反応様式により5つのクループに分類されている、異性体間の転換を触媒するEC5群の酵素の総称。
10. リン酸を含み、生物細胞の種々の膜系、例えば原形質膜、核膜、リソソーム膜などを構成する主要な脂質。

| | | |
|---|---|---|
| 影響 | エイキョウ | influence |
| 戻る | もどる | to return, to revert |
| 元来 | ガンライ | naturely, innately |
| 連鎖 | レンサ | linkage |
| 比較 | ヒカク | comparison |
| 偶然の | グウゼンの | accidental, due to chance |
| 受粉 | ジュフン | pollination |
| 単鎖 | タンサ | single chain |
| 置き換わる | おきかわる | to be replaced |
| 完全 | カンゼン | completeness |
| 失活 | シッカツ | inactivation |
| 野生型 | ヤセイがた | wild type |
| 転換 | テンカン | conversion |

**Ex. 5.6 Additional sentence translations**

1. 免疫電子顕微鏡法では抗原抗体反応の特異性を利用して構造体内の蛋白質分子の配置を知ることができる。精製された小型ウイルスなどの場合、ウイルスの各成分を精製し、それぞれに対する免疫抗体をつくり、特異な抗体をウイルスと反応させ、抗体分子の所在を負染色法により観察する。大きな細胞内部の構造体の場合、フェリチンで標識した抗体を用い、フェリチン内の鉄の分布を切片法により調べる。

2．蛍光抗体法では組織標本中の酵素、構造蛋白質、微生物などの抗原生物質を、蛍光標識抗体と結合させ、蛍光顕微鏡下に検出する。検出すべき抗原に対する抗体を直接蛍光標識する方法のほかに、抗原に対する抗体に対する抗体つまり第二抗体を標識する方法(間接法)、補体を介して第二蛍光標識抗体を結合させる方法などがある。

3．紫外線損傷を受けた生物体が修復する過程には光を必要とするもの(光回復)としないもの(暗修復)がある。前者は光回復酵素によるもので、ピリミジン二量体にのみ有効であるが、後者は広い範囲のDNA損傷に働く。暗修復には除去修復、複製後修復、SOS修復およびアルキル化損傷の適応修復があり、いずれも酵素反応の複雑な連鎖から成る損傷DNAの修復機構である。

4．突然変異によって変化を受けたDNAのヌクレオチドの配列が完全に元の状態に戻った場合は真の復帰突然変異である。突然変異が起こって変化ししたヌクレオチド配列の近くに第二の突然変異が起こることによって、遺伝子型は元に戻らなくても表現形質が野生型に戻ることがある。フレームシフト変異の場合には、最初の突然変異を起こした物質と同じタイプのフレームシフトを起こす突然変異誘発物質によって復帰突然変異が起こるが、違うタイプの突然変異誘発物質では復帰突然変異は起こらない。

5．移植片の提供者とその受容者の間の遺伝的背景が異なっている場合、当然移植片の抗原に対して受容者による一連の免疫応答が引き起こされ最終的には移植片の壊死脱落によって拒絶反応が完結する。この免疫応答は細胞性と液性(抗体性)の反応が複雑に絡み合った過程であり、時期的に超急性拒絶反応、急性拒絶反応、慢性拒絶反応などに大別される。

6．温度感受性変異株の中にはある温度以上で野生株と異なる表現型を示す高温感受性変異株と、ある温度以下で野生株と異なる表現型を示す低温感受性変異株とがある。一般には前者は変異した遺伝子の産物である蛋白質あるいはRNAがある温度以上で不安定化ないし失活するために、その表現型が高温感受性になる場合であり、一方、後者はリボソームで典型的に示されているように、細胞内構造体の構成蛋白質に変異が生じた結果、一定温度以下では構造体の形成が起こらなくなるような場合である。

7．血液は、血管内を循環している間は流動性であるが、一旦血管外に取り出されると、やがて流動性を失う。これは、血漿中のフィブリノーゲンが繊維状のフィブリンに転換し、フィブリン網を形成し、血液全体がゲル状になるためである。この現象を血液凝固という。血液凝固の結果、形成された血液凝塊は、フィブリン網の間に、赤血球、白血球などの血球成分を取り込んでいる。

8．α-1,4-グルカン分枝酵素が澱粉系多糖のα-1,4-グルカン鎖の一部を6位に転移して分枝をつくり、アミロース型(直鎖型(多糖からアミロペクチン-グリコーゲン型の分子多糖をつくる。この酵素は動物、植物に広く存在し、植物のものはQ酵素と呼ばれる。活性はアミロースまたは可溶性澱粉に作用させて、そのヨウ素-澱粉反応の強度の現象を測定するか、あるいはホスホリラーゼによるグルコース1-リン酸からの多糖合成に際して本酵素を添加し、反応速度の増加を無添加の場合と比較して測定する。

9．ワトソン-クリックのモデルで、もっとも驚くべきことは、二本のDNAの鎖がらせん構造をしているということだけではなく、二本の鎖をつくるヌクレオチドの配列、つまりは塩基の配列に相補的な性質があるということであった。具体的にいうと、片方の鎖にAがあると、他方の鎖の相当する位置にはTがあり、GがあるとCがあるということである。したがって、片方の鎖の塩基配列が、例えばACGTであれば、他方の鎖の相当する位置の塩基配列はTGCAとなる。これは、AとT、CとGのそれぞれの分子間に、うまく結合することのできる部分があるからである。

10．放射線や化学物質、あるいは他の生物体、生体物質などが細胞に障害を与える場合、そのような特性を細胞毒性という。障害の種類としては細胞死、増殖能、DNA合成能を含む細胞の各種機能の低下がある。化学物質や物理的要因による障害を細胞毒性、生物体による障害を細胞障害という傾向がある。細胞にはこれらの障害をある程度修復する能力がある場合がある。

| | | |
|---|---|---|
| 精製 | セイセイ | purification |
| 小型 | こがた | small in size |
| 所在 | ショザイ | location |
| 負染色 | フセンショク | negative staining |
| 観察 | カンサツ | observation |
| 切片法 | セッペンホウ | microtomy |
| 標本 | ヒョウホン | specimen |
| 範囲 | ハンイ | extent, scope |
| 適応 | テキオウ | adaptation |
| 真の | シンの | true, genuine |
| 突然変異誘発物質 | トツゼンヘンイユウハツブッシツ | mutagen |
| 提供者 | テイキョウシャ | donor |
| 背景 | ハイケイ | background |
| 当然 | トウゼン | naturally |
| 免疫応答 | メンエキオウトウ | immune response |
| 引き起こす | ひきおこす | to bring about, to cause |
| 最終的に | サイシュウテキに | finally |
| 壊死脱落 | エシダツラク | necrosis and peeling |
| 拒絶反応 | キョゼツハンノウ | rejection reaction |
| 完結 | カンケツ | completion, conclusion |
| 絡み合う | からみあう | to be intertwined |
| 超急性 | チョウキュウセイ | hyperacute |
| 急性 | キュウセイ | acute |
| 慢性 | マンセイ | chronic |
| 典型的に | テンケイテキに | representatively |
| 循環 | ジュンカン | circulation |
| 一旦 | イッタン | once |
| 取り出す | とりだす | to take out |
| 繊維 | センイ | fiber |
| 網 | モウ | network |
| 取り込む | とりこむ | to incorporate to take in |
| 澱粉 | デンプン | starch |
| 転移 | テンイ | rearrangement, transition, dislocation, metastasis |
| 本- | ホン | this … |
| 具体的に | グタイテキに | concretely, definitely |
| うまく | | successfully, smoothly, well |
| -死 | シ | death |
| 傾向 | ケイコウ | tendency |

**Translations for Ex. 5.4**

1. The term "cell strain" indicates a cultured line that has come by means of selection or cloning to possess a specific characteristic or (genetic) label, either from the first generation culture or from the cell line.
2. At the time of infection by phage, we call the number of phage particles added per bacterial cell the multiplicity of infection. When the multiplicity is *m*, the proportion $P_r$ of bacteria that are infected by *r* phage particles is given by $P_r=(m^r)(e^{-m})/(r!)$.
3. Plasma clots can be obtained from blood whose clotting has been inhibited by citric acid or heparin. Plasma obtained after removal of the blood cells via centrifugation is mixed with liquid (in a simple case a liquid such as tissue extract) that contains a clotting factor
4. The migration of a branch in a figure-eight-shaped DNA molecule is required in order to form a long heteroduplex portion. Thus, it is thought that branch migration is an indispensable process in general recombination. Also, in the explanation of gene conversion by the Holliday model branch migration is taken as an essential condition for the form of the recombination intermediate.
5. If the various conditions in a gyratory culture are held constant, a cell aggregate is formed with fixed size and shape and extremely high reproducibility. For this reason gyratory cultures are widely used in research on cell aggregation and cell sorting
6. Among the toxins produced by cells exotoxins are polymeric proteins and are unstable with respect to heat. Their toxicity is extremely high. The classical diptheria toxin, the tetanus toxin and the botulinum toxin are examples.
7. For a chemotherapeutic agent it is desirable not to cause injury to humans or to useful species such as domestic animals, but to exhibit selective toxicity only toward harmful species such as parasitic insects or infectious or pathogenic microbes.
8. Epidermal growth factors (EGF) are chemical substances that act on various types of epidermal tissue and selectively promote their growth. They resemble nerve growth factors (NGF) in that they are secreted by the salivary glands, but these EGF do not bind to insulin receptors or nerve growth factor receptors
9. All diseases are thought to arise from a complicated intertwinement of hereditary factors and environmental factors. Those diseases for which hereditary factors are directly linked to the expression of the disease are called hereditary diseases.
10. For many animals the [development of the] oocyte comes to a halt at the first mature prephase of cell division until ovulation. In this state the oocyte possesses a huge nucleus. This prephase is an extremely long process, and during this period the synthesis of mRNA flourishes.

| | | |
|---|---|---|
| 緩 | カン | relaxation; mitigation |
| 鎖 | サ | chain, strand |
| | くさり | chain |
| 指 | シ | indication; finger |
| | さ(す) | to indicate, point |
| | ゆび | finger |
| 識 | シキ | knowledge, recognition |
| 終 | シュウ | end |
| | お(える) | to end, complete {v.t.} |
| | おわ(る) | to end {v.t., v.i.} |
| | おわ(り) | end, conclusion |
| 親 | シン | intimacy; parent |
| | おや | parent |
| 走 | ソウ | running |
| | はし(る) | to run |
| 泌 | ヒ;ヒツ | secretion |
| 遊 | ユウ | idle, wandering |
| 抑 | ヨク | suppression, restraint |
| | おさ(える) | to suppress, restrain |

緩 鎖 指 識 終 親 走 泌 遊 抑

## 緩

| | | |
|---|---|---|
| 核緩和 | カクカンワ | nuclear relaxation |
| 緩衝液 ★ | カンショウエキ | buffer |
| 緩衝作用 | カンショウサヨウ | buffering action |
| 緩衝指数 | カンショウシスウ | buffer index |
| 緩衝値 | カンショウチ | buffer value |
| 緩和 ★ | カンワ | relaxation |
| 緩和時間 | カンワジカン | relaxation time |
| 緩和法 | カンワホウ | relaxation method |
| 縦緩和時間 | たてカンワジカン | longitudinal relaxation time |
| 弛緩 ★ | シカン | relaxation |
| 弛緩因子 | シカンインシ | relaxation factor |
| 弛緩液 | シカンエキ | relaxation solution |
| 弛緩性麻痺 | シカンセイマヒ | flaccid paralysis |
| 弛緩熱 | シカンネツ | heat of relaxation |
| 横緩和時間 | よこカンワジカン | transverse relaxation time |
| 両性イオン緩衝液 | リョウセイイオンカンショウエキ | zwitterionic buffer |

## 鎖

| | | |
|---|---|---|
| 一重鎖RNA | イチジュウサRNA | single-stranded RNA |
| 一本鎖RNA | イッポンサRNA | single-stranded RNA |
| 核鎖繊維 | カクサセンイ | nuclear chain fiber |
| 軽鎖 | ケイサ | light chain |
| 呼吸鎖 | コキュウサ | respiratory chain |
| 鎖延長因子 | サエンチョウインシ | chain elongation factor |
| 鎖終結因子 | サシュウケツインシ | chain-releasing factor |
| 三本鎖ヘリックス | サンボンサヘリックス | triple-stranded helix |
| 重鎖 | ジュウサ | heavy chain |
| 情報鎖 | ジョウホウサ | formative strand |
| 相補鎖 ★ | ソウホサ | complementary chain |
| 側鎖 | ソクサ | side chain |
| 側鎖説 | ソクサセツ | side chain theory |
| 短鎖 | タンサ | short chain |
| 長鎖塩基 | チョウサエンキ | long chain base |
| 直鎖 | チョクサ | normal (straight) chain |
| 糖鎖 | トウサ | sugar chain |
| 糖鎖形成 | トウサケイセイ | glycosylation |
| 糖鎖不全説 | トウサフゼンセツ | loss of extension of glycolipid sugar chain |
| 二本鎖RNA ★ | ニホンサRNA | double-stranded RNA |

| | | |
|---|---|---|
| 二本鎖RNA依存性プロテインキナーゼ | ニホンサRNAイゾンセイプロテインキナーゼ | double-stranded RNA-dependent protein kinase |
| 二本鎖RNAウイルス | ニホンサRNAウイルス | double-stranded RNA virus |
| 閉鎖 | ヘイサ | atresia |
| 閉鎖コロニー | ヘイサコロニー | closed colony |
| 溶血性連鎖[状]球菌 | ヨウケツセイレンサ[ジョウ]キュウキン | *Streptococcus haemolyticus* |
| 連鎖 ★ | レンサ | linkage |
| 連鎖形質導入 | レンサケイシツドウニュウ | linked transduction |
| 連鎖[状]球菌 | レンサ[ジョウ]キュウキン | *Streptococcus* |
| 連鎖地図 | レンサチズ | linkage map |

# 指

| | | |
|---|---|---|
| 緩衝指数 | カンショウシスウ | buffer index |
| 細胞毒性指数 | サイボウドクセイシスウ | cytotoxic index |
| 酸性度指数 | サンセイドシスウ | acidity index |
| 指示菌 ★ | シジキン | indicator bacterium |
| 指示薬 | シジヤク | indicator |
| 指数 ★ | シスウ | index |
| 指数増殖期 | シスウゾウショクキ | exponential growth phase |
| 指定 | シテイ | designation, assignment |
| 指標 ★ | シヒョウ | indicator |
| 指標酵素 | シヒョウコウソ | marker enzyme |
| 指標生物 | シヒョウセイブツ | indicator (organism) |
| 指紋 | シモン | dactylogram |
| 指紋法 | シモンホウ | fingerprinting (method) |
| 指令 | シレイ | order, instruction |
| 水素指数 | スイソシスウ | hydrogen exponent |
| 分裂指数 | ブンレツシスウ | mitotic index |
| 遊走指数 | ユウソウシスウ | migration index |

# 識

| | | |
|---|---|---|
| 遺伝標識形質 ★ | イデンヒョウシキケイシツ | genetic marker |
| 抗原識別部 | コウゲンシキベツブ | antigen-recognition site |
| 後標識 | コウヒョウシキ | postlabeling |
| 細胞認識 | サイボウニンシキ | cell recognition |
| 識別 | シキベツ | identification, discernment |
| 瞬間標識 | シュンカンヒョウシキ | pulse-labeling |
| 瞬間標識追跡実験 | シュンカンヒョウシキツイセキジッケン | pulse-chase experiment |
| 親和性標識 | シンワセイヒョウシキ | affinity labeling |

| | | |
|---|---|---|
| 前標識 | ゼンヒョウシキ | prelabeling |
| 多重標識 | タジュウヒョウシキ | multiple label |
| 認識 ★ | ニンシキ | recognition |
| 標識 ★ | ヒョウシキ | label(ing) |
| 標識遺伝子 | ヒョウシキイデンシ | marker gene |
| 標識化 | ヒョウシキカ | labeling |
| 標識化合物 | ヒョウシキカゴウブツ | labeled compound |
| 標識法 | ヒョウシキホウ | labeling |
| 放射能標識 | ホウシャノウヒョウシキ | radiolabeling |

## 終

| | | |
|---|---|---|
| 最終 | サイシュウ | final |
| 最終産物 | サイシュウサンブツ | end product |
| 最終産物阻害 | サイシュウサンブツソガイ | end product inhibition, feedback inhibition |
| 最終産物抑制 | サイシュウサンブツヨクセイ | end product repression, feedback repression |
| 最終段階 | サイシュウダンカイ | culmination |
| 最終分化細胞 | サイシュウブンカサイボウ | terminally differentiated cell |
| 鎖終結因子 | サシュウケツインシ | chain-releasing factor |
| 終期 ★ | シュウキ | telophase |
| 終結因子 | シュウケツインシ | release factor |
| 終止コドン ★ | シュウシコドン | termination codon |
| 終脳 | シュウノウ | telencephalon |
| 終了因子 ★ | シュウリョウインシ | termination factor |
| 分裂終了細胞 | ブンレツシュウリョウサイボウ | postmitotic cell |

## 親

| | | |
|---|---|---|
| 親株 | おやかぶ | parent strain |
| 親細胞 ★ | おやサイボウ | parent cell |
| 基質親和性 | キシツシンワセイ | substrate affinity |
| 近親交配 | キンシンコウハイ | inbreeding |
| 親液性原子団 | シンエキセイゲンシダン | lyophilic |
| 親水基 | シンスイキ | hydrophilic group |
| 親水性アミノ酸 | シンスイセイアミノサン | hydrophilic amino acid |
| 親和性標識 ★ | シンワセイヒョウシキ | affinity labeling |
| 親和溶出 | シンワヨウシュツ | affinity elution |
| 親和力 ★ | シンワリョク | affinity |
| 臓器親和性 | ゾウキシンワセイ | organotroph |
| 電子親和力 | デンシシンワリョク | electron affinity |
| 母親の | ははおやの | maternal |
| 免疫親和性 | メンエキシンワセイ | immunoaffinity |

# 走

| | | |
|---|---|---|
| 示差走査熱量測定法 | シサソウサネツリョウソクテイホウ | differential scanning calorimetry |
| 走圧性 | ソウアツセイ | barotaxis |
| 走化性 ★ | ソウカセイ | chemotaxis |
| 走化性因子 | ソウカセイインシ | chemotactic factor |
| 走光性 | ソウコウセイ | phototaxis |
| 走査 ★ | ソウサ | scanning |
| 走査型電子顕微鏡 | ソウサがたデンシケンビキョウ | scanning electron microscope |
| 走査型トンネル顕微鏡 | ソウサがたトンネルケンビキョウ | scanning tunneling microscope |
| 単球走化[性]因子 | タンキュウソウカ[セイ]インシ | monocyte chemotactic factor |
| 白血球遊走促進因子 | ハッケッキュウユウソウソクシンインシ | leukocyte migration enhancement factor |
| 白血球遊走阻止因子 | ハッケッキュウユウソウソシインシ | leukocyte migration inhibition factor |
| 迷走神経 | メイソウシンケイ | vagus (nerve) |
| 遊走 ★ | ユウソウ | migration, wandering |
| 遊走子 | ユウソウシ | zoospore |
| 遊走子形成菌類 | ユウソウシケイセイキンルイ | zoosporic fungi |
| 遊走指数 | ユウソウシスウ | migration index |
| 遊走阻止因子 | ユウソウソシインシ | migration inhibition factor |

# 泌

| | | |
|---|---|---|
| 外分泌 | ガイブンピ[ツ] | external secretion |
| 神経内分泌 | シンケイナイブンピ[ツ] | neuroendocrine |
| 神経分泌 ★ | シンケイブンピ[ツ] | neurosecretion |
| 神経分泌細胞 | シンケイブンピ[ツ]サイボウ | neurosecretory cell |
| 内分泌 | ナイブンピ[ツ] | internal secretion |
| 内分泌学 ★ | ナイブンピ[ツ]ガク | endocrinology |
| 泌乳 | ヒ[ツ]ニュウ | lactation |
| 泌乳刺激ホルモン | ヒ[ツ]ニュウシゲキホルモン | lactogenic hormone |
| 分泌 ★ | ブンピ[ツ] | secretion |
| 分泌顆粒 | ブンピ[ツ]カリュウ | secretory granule |
| 分泌細胞 | ブンピ[ツ]サイボウ | secretory cell |
| 分泌促進物質 | ブンピ[ツ]ソクシンブッシツ | secretagogue |
| 分泌蛋白質 | ブンピ[ツ]タンパクシツ | secretory protein |
| 分泌物 | ブンピ[ツ]ブツ | secrete |
| 分泌ベクター | ブンピ[ツ]ベクター | secretion vector |
| 分泌片 | ブンピ[ツ]ヘン | secretory piece |

## 遊

| | | |
|---|---|---|
| 寒天内浮遊培養法 | カンテンナイフユウバイヨウホウ | agar suspension culture |
| 白血球遊走促進因子 | ハッケッキュウユウソウソクシンインシ | leukocyte migration enhancement factor |
| 白血球遊走阻止因子 | ハッケッキュウユウソウソシインシ | leukocyte migration inhibition factor |
| 浮遊[細胞]培養 ★ | フユウ[サイボウ]バイヨウ | suspension (cell) culture |
| 浮遊性 | フユウセイ | floating |
| 遊走 ★ | ユウソウ | migration, wandering |
| 遊走子 | ユウソウシ | zoospore |
| 遊走子形成菌類 | ユウソウシケイセイキンルイ | zoosporic fungi |
| 遊走指数 | ユウソウシスウ | migration index |
| 遊走阻止因子 | ユウソウソシインシ | migration inhibition factor |
| 遊離 | ユウリ | free |
| 遊離型ロボソーム | ユウリがたリボソーム | free ribosome |
| 遊離基 ★ | ユウリキ | free radical |
| 遊離細胞 | ユウリサイボウ | free cell |
| 遊離水 | ユウリスイ | free water |

## 抑

| | | |
|---|---|---|
| 胃抑制ポリペプチド | イヨクセイポリペプチド | gastric inhibitory polypeptide |
| 組込み抑制 | くみこみヨクセイ | integrative suppression |
| 酵素抑制 | コウソヨクセイ | enzyme repression |
| 最終産物抑制 | サイシュウサンブツヨクセイ | end product repression |
| 成長抑制剤 | セイチョウヨクセイザイ | growth retardant |
| 接触抑制 ★ | セッショクヨクセイ | contact inhibition |
| 補抑制物質 | ホヨクセイブッシツ | corepressor |
| 免疫抑制 ★ | メンエキヨクセイ | immunosuppression |
| 免疫抑制遺伝子 | メンエキヨクセイイデンシ | immune suppression gene |
| 免疫抑制剤 | メンエキヨクセイザイ | immunosuppressive agent |
| 抑制 ★ | ヨクセイ | depression, repression, suppression, inhibition |
| 抑制因子 | ヨクセイインシ | inhibitor, repressor |
| 抑制酵素 | ヨクセイコウソ | repressible enzyme |
| 抑制性シナプス | ヨクセイセイシナプス | inhibitory synapse |
| 抑制性シナプス後電位 | ヨクセイセイシナプスゴデンイ | inhibitory postsynaptic potential |
| 抑制性ニューロン | ヨクセイセイニューロン | inhibitory neuron |
| 抑制的対立遺伝子 | ヨクセイテキタイリツイデンシ | antimorph |

*SUPPLEMENTARY VOCABULARY USING KANJI FROM CHAPTER 16*

| | | |
|---|---|---|
| 暗号 | アンゴウ | (genetic) code |
| 遺伝暗号 | イデンアンゴウ | genetic code, codon |
| 遺伝距離 | イデンキョリ | genetic distance |
| 遺伝子発現 | イデンシハツゲン | gene expression |
| 環状DNA | カンジョウDNA | circular DNA |
| 酵素の特異性 | コウソのトクイセイ | enzyme specificity |
| 条件 | ジョウケン | condition |
| 制御 | セイギョ | regulation, control |
| 制限 | セイゲン | restriction |
| 制限酵素 | セイゲンコウソ | restriction enzyme |
| 組織特異性 | ソシキトクイセイ | tissue specificity |
| 炭素循環 | タンソジュンカン | carbon cycle |
| 窒素循環 | チッソジュンカン | nitrogen cycle |
| 中和 | チュウワ | neutralization |
| 特異性 | トクイセイ | specificity |
| 特殊形質導入 | トクシュケイシツドウニュウ | specialized transduction |
| 発現 | ハツゲン | expression, manifestation |
| 表現 | ヒョウゲン | expression |
| 表現度 | ヒョウゲンド | expressivity |
| 標的器官 | ヒョウテキキカン | target organ |
| 標本 | ヒョウホン | specimen |
| 負の制御 | フのセイギョ | negative control |
| 不飽和 | フホウワ | unsaturation |
| 不和合性 | フワゴウセイ | incompatibility |
| 分離 | ブンリ | segregation |
| 分離の法則 | ブンリのホウソク | law of segregation |
| 閉環状DNA | ヘイカンジョウDNA | closed circular DNA |
| 紡錘体 | ボウスイタイ | spindle |
| 飽和 | ホウワ | saturation |
| 連結酵素 | レンケツコウソ | ligase |
| 連結反応 | レンケツハンノウ | ligation |

# EXERCISES

### Ex. 6.1 Matching Japanese and English terms

| | | |
|---|---|---|
| (　) 遺伝標識形質 | (　) 最終産物抑制 | (　) 分泌細胞 |
| (　) 親株 | (　) 鎖終結因子 | (　) 免疫抑制遺伝子 |
| (　) 緩衝指数 | (　) 指標酵素 | (　) 遊走指数 |
| (　) 寒天内浮遊培養法 | (　) 神経分泌 | (　) 遊走阻止因子 |
| (　) 緩和時間 | (　) 親和性標識 | (　) 連鎖形質導入 |

| | | |
|---|---|---|
| 1. affinity labeling | 6. genetic marker | 11. migration inhibition factor |
| 2. agar suspension culture | 7. immune suppression gene | 12. neurosecretion |
| 3. buffer index | 8. linked transduction | 13. parent strain |
| 4. chain-releasing factor | 9. marker enzyme | 14. relaxation time |
| 5. end product repression | 10. migration index | 15. secretory cell |

### Ex. 6.2 KANJI with similar structural elements

Look carefully at each of the two KANJI on the left, and note which structural element is common to both. Combine each KANJI on the left with the appropriate KANJI on the right to make a meaningful JUKUGO. Each technical term that contains one or more of the 100 KANJI introduced in this book can be found in the vocabulary lists for those KANJI. Other terms can be found in one of the supplementary vocabulary lists, including Lesson 0.

| | | | |
|---|---|---|---|
| 1. | (1) 還　(2) 環 | 酸化( )元 | 炭素循( ) |
| 2. | (1) 緩　(2) 終 | ( )衝作用 | ( )了因子 |
| 3. | (1) 凝　(2) 旋 | 右( )性 | ( )集 |
| 4. | (1) 指　(2) 脂 | ( )示菌 | リン( )質 |
| 5. | (1) 指　(2) 抑 | 酵素( )制 | ( )標生物 |
| 6. | (1) 識　(2) 織 | 組( )特異性 | 標( )遺伝子 |
| 7. | (1) 旋　(2) 族 | ( )光計 | 同( )体 |
| 8. | (1) 毒　(2) 母 | 細胞( ) | 卵( )細胞 |
| 9. | (1) 泌　(2) 泳 | 電気( )動 | 内分( )学 |
| 10. | (1) 病　(2) 疫 | 遺伝( ) | 免( )親和性 |
| 11. | (1) 遊　(2) 遠 | 高速( )心機 | 浮( )培養 |
| 12. | (1) 抑　(2) 御 | 制( ) | ( )制因子 |

### Ex. 6.3 Matching Japanese technical terms with definitions

Read each definition carefully, and then choose the appropriate technical term. Words that you have not yet encountered are listed following the definitions.

( ) 緩衝液
( ) 呼吸鎖
( ) 指示菌
( ) 終期
( ) 親株
( ) 標識化合物
( ) 分泌蛋白質
( ) 遊走阻止因子
( ) 遊離基
( ) 抑制酵素

1. 弱酸とその塩、あるいは弱塩基とその塩の混合溶液。
2. 酸化的リン酸化反応において酸化還元をつかさどる酵素系。
3. ファージの力価や平板効率を測定したり、特定のファージを他のファージと区別、同定する目的で用いられる細菌株。
4. 化合物の構成元素の一部を、放射性同位体や安定同位体で置き換えた化合物。
5. 細胞分裂の過程の最後の時期。
6. 細胞融合、核移植などの細胞再構成法を用いて複数の細胞系統から一つの細胞系統を得た時、その元となった細胞系統。
7. 抗原と反応したリンパ球の生産する物質で、マクロファージに作用してその運動を阻止する活性をもつもの。
8. 細胞内で合成され、細胞膜外へ分泌される蛋白質。
9. 不対電子を有する化学種。
10. 特定の代謝物質の細胞内濃度が増加すると合成率が減少する酵素。

| | | |
|---|---|---|
| 弱酸 | ジャクサン | weak acid |
| 弱塩基 | ジャクエンキ | weak base |
| つかさどる | | to manage, direct |
| 力価 | リキカ | titer |
| 平板効率 | ヘイバンコウリツ | plating efficiency |
| 時期 | ジキ | time, period |
| 複数 | フクスウ | two or more, several |

**Ex. 6.4 Sentence translations**

Read each sentence carefully, and then translate it. Words that you have not yet encountered are listed following the sentences.

1. 緩和という現象には、注目する系と格子系との動的相互作用が反映されているので、注目する系の動的情報が緩和時間の測定から得られることは重要である。
2. RNAウイルスの内、ポリオウイルスやレトロウイルスなどではゲノムは一本鎖RNAであるが、レオウイルスなどでは二本鎖RNAをゲノムとしてもつ。
3. 最初の細胞数をaとし、t時間内にx個の細胞が分裂して総数がb個の細胞になった時、組織または細胞集団中の分裂細胞の頻度を示す分裂指数は次の式で表される。
4. 遺伝分析では雑種形成実験を行って、ある遺伝形質に関係する遺伝子の数、性質、連鎖関係を決定するが、その分析の基礎は標識遺伝子の交雑実験における挙動を追跡することである。
5. 生体は過剰の産物が蓄積しないように、代謝制御を行っている。つまり、一連の代謝経路において、その経路上にある酵素(主に初段階の酵素)が最終産物により阻害を受ける。
6. 酵素分子には、基質結合部位といわれるその酵素に特異的な基質を結合する部位があり、基質親和性はこの部位と基質分子との結合の強さの尺度である。
7. 細菌が栄養物に集まる行動(正の走化性(や酸やアルカリから逃避する行動(負の走化性(をはじめとして、発生分化の過程で見られる細胞の一定部位への移動など細胞の合目的な移動の多くが走化性に基づいている。
8. 細胞外へ分泌される遺伝子産物の多くの場合はその産物の前駆体のN末端領域に数十アミノ酸からなる、分泌のためのシグナル配列が存在する。前駆体が細胞膜通過の際、シグナル配列が切断され、産物が成熟型として分泌される。
9. 浮遊培養では足場依存性を示す細胞は増殖しない。しかし、このような細胞では浮遊性の微粒子に細胞を付着させてからかくはん培養する方法によって浮遊培養ができる。
10. 酵素阻害が酵素の分子数を変えずに酵素活性を阻害する制御であるのに対し、酵素抑制は酵素の分子数のみを低下させる制御である。

| | | |
|---|---|---|
| 格子系 | コウシケイ | lattice |
| 相互作用 | ソウゴサヨウ | interaction |
| 反映 | ハンエイ | reflection |
| 注目する | チュウモクする | to pay attention |
| 総数 | ソウスウ | total count |
| 基礎 | キソ | foundation, basis |
| 挙動 | キョドウ | action, behavior |
| 追跡 | ツイセキ | tracking |
| 過剰 | カジョウ | excess |
| 蓄積 | チクセキ | accumulation |
| 初段階 | ショダンカイ | initial stage |
| 尺度 | シャクド | measure, scale |
| 逃避 | トウヒ | evasion, escape |
| 負の | フの | negative |
| 合目的な | ゴウモクテキな | purposeful |
| 末端領域 | マッタン リョウイキ | distal region |
| 切断 | セツダン | cleavage |
| 成熟型 | セイジュクがた | mature form |
| 足場 | あしば | anchorage |
| 依存性 | イゾンセイ | dependency |

**Ex. 6.5 Additional dictionary entries**

( ) 遺伝暗号
( ) 遺伝距離
( ) 遺伝子発現
( ) 環状DNA
( ) 基質特異性
( ) 制限酵素
( ) 組織特異性
( ) 特殊形質導入
( ) 発現
( ) 閉環状DNA

1. 遺伝子によって決定される形質が表現型として現れてくること。
2. 遺伝物質がDNAの塩基配列という形で保存されていたため、元々存在しなかったり隠れていたりする機能や特徴が現れ出ること。
3. 多くのプラスミドや増殖中のファージDNAに見られる、まったく切れ目のない二本鎖の環状DNA。
4. 種間、亜種間など、一般に異なった生物集団間の遺伝的差異を遺伝子頻度や相同なDNAの塩基配列の相違を用いて表した尺度。
5. 組織に固有の構成成分の細胞の性質、またはその細胞が生産する細胞間物質の性質に由来する各組織が他の組織と区別されること。
6. ただ一種類の相手物質、つまり基質、が酵素の触媒作用を受て化学反応を起こすという酵素のもつ性質。
7. 蛋白質のアミノ酸配列を規定するための情報をもつ核酸塩基の配列。
8. DNAの両末端が共有結合によって連結された状態で、超コイルのない開環状、負の超コイルをもつ閉環状、または正の超コイルをもつ閉環状の三つの状態を取り得るもの。
9. 特定の配列の塩基対が存在するときだけ、そこでDNAの二重らせんを切断する酵素。
10. ファージにより形質導入し得る宿主菌の遺伝物質がある特定の形質に限られる現象。

| | | |
|---|---|---|
| 元々 | もともと | originally |
| 隠れる | かくれる | to be concealed |
| 切れ目 | きれめ | nick |
| 相違 | ソウイ | difference, disparity |
| 相手物質 | あいてブッシツ | matching substance |
| 規定 | キテイ | regulation, provision |
| 超コイル- | チョウコイル- | supercoiled |
| 開環状- | カイカンジョウ- | open-circular |
| 閉環状- | ヘイカンジョウ- | closed-circular |
| 取る | とる | to take, to get |

**Translations for Ex. 6.4**

1. The dynamic interaction between the system we are observing and the lattice is reflected in the phenomenon of relaxation. Therefore, the fact that we can from measurement of the relaxation time obtain dynamic information about the system we are observing is important.
2. Among RNA viruses the genomes of viruses such as the polio virus and retroviruses are single-stranded RNA. However, viruses such as reoviruses possess double-stranded RNA for their genomes.
3. When the initial number of cells is $a$ and in time $t$ a number $x$ of those cells have divided so that the total count of cells has become $b$, the mitotic index, which indicates the frequency of divided cells within the tissue or cell population, is expressed by the following equation.
4. In genetic analysis we carry out hybridization experiments to determine the number, characteristics and linkage relationships of the genes related to a certain genetic trait. The basis for that analysis lies in tracking the behavior of marker genes during crossing experiments.
5. Organisms carry out metabolic regulation so that excess metabolic products will not accumulate. That is, in a series of metabolic pathways enzymes (principally the enzymes in the initial stages) that exist in those pathways are inhibited by the end products.
6. In an enzyme molecule there is a site that binds to the substrate that is specific to that enzyme. The site is called the substrate-binding site. Substrate affinity is a measure of the strength of the binding between this site and the substrate molecule.
7. Many of the purposeful movements of cells, such as movement to a determined position during the processes of development and differentiation, not to mention the behavior of bacteria in gathering around nutrients (positive chemotaxis) or escaping from acids or alkali (negative chemotaxis), are based on chemotaxis.
8. For many of the gene products secreted outside the cell there exist on the N distal region of the product precursors so-called "signal sequences," which are composed of several dozen amino acids and are [signals] for secretion [of those products]. When the precursors pass through the cell membrane, the signal sequences are cleaved off, and the products are secreted in their mature forms.
9. Cells that display anchorage dependency will not grow in a suspension culture. However, cells of this type can be suspension cultured by first attaching the cells to minute floating particles, and then carrying out a spinner culture.
10. Enzyme inhibition is [a type of] control in which we do not change the number of enzyme molecules; we inhibit the activity of the enzyme. In contrast, enzyme repression is [a type of] control in which we reduce only the number of enzyme molecules.

| | | |
|---|---|---|
| 画 | ガ | picture, image |
| | カク | mark, demarcation; stroke in a character |
| 拡 | カク | spreading, expanding |
| 群 | グン | group, herd |
| | む(れ) | group, crowd |
| 欠 | ケツ | lack, defect |
| | か(く) | to lack |
| | か(ける) | to be missing |
| 座 | ザ | seat |
| 始 | シ | beginning, starting |
| | はじ(める) | to begin, initiate |
| | はじ(まる) | to begin, open |
| | はじ(め) | beginning, start |
| 写 | シャ | copying, projecting |
| | うつ(す) | to copy, photograph |
| | うつ(る) | to be projected, photographed |
| 症 | ショウ | disease; symptom |
| 答 | トウ | answer, response |
| | こた(える) | to answer, respond |
| | こた(え) | answer, response |
| 頻 | ヒン | repeating |

画 拡
群 欠
座 始
写 症
答 頻

# 画

| | | |
|---|---|---|
| 画線　★ | カクセン | streak |
| 画線培養 | カクセンバイヨウ | streak culture |
| 区画　★ | クカク | compartment |
| 交差画線培養 | コウサカクセンバイヨウ | cross streak culture |
| 細胞内区画 | サイボウナイクカク | intracellular compartmentation |
| 細胞分画法 | サイボウブンカクホウ | cell fractionation |
| 分画　★ | ブンカク | fraction; fractionation |
| 分画遠心分離 | ブンカクエンシンブンリ | differential centrifugation |
| 硫安分画 | リュウアンブンカク | ammonium sulfate fractionation |

# 拡

| | | |
|---|---|---|
| 回転拡散係数 | カイテンカクサンケイスウ | rotary diffusion coefficient |
| 拡散　★ | カクサン | diffusion |
| 拡散因子 | カクサンインシ | spreading factor |
| 拡散近似 | カクサンキンジ | diffusion approximation |
| 拡散係数 | カクサンケイスウ | diffusion coefficient |
| 拡散電位 | カクサンデンイ | diffusion potential |
| 拡張　★ | カクチョウ | expansion, dilation |
| 血管内皮由来拡張因子 | ケッカンナイヒユライカクチョウインシ | endothelium-derived relaxant factor |
| 最大拡張期電位 | サイダイカクチョウキデンイ | maximum diastolic potential |
| 自由拡散 | ジユウカクサン | free diffusion |
| 促進拡散 | ソクシンカクサン | facilitated diffusion |
| 側方拡散 | ソクホウカクサン | lateral diffusion |
| 単純放射状免疫拡散 | タンジュンホウシャジョウメンエキカクサン | single radial immunodiffusion |
| 二重拡散法 | ニジュウカクサンホウ | double diffusion test |
| 二重免疫拡散法 | ニジュウメンエキカクサンホウ | double immunodiffusion |
| 微量拡散法 | ビリョウカクサンホウ | microdiffusion method |
| 放射状免疫拡散 | ホウシャジョウメンエキカクサン | radial immunodiffusion |
| 免疫拡散法　★ | メンエキカクサンホウ | immunodiffusion |
| 免疫電気拡散法 | メンエキデンキカクサンホウ | immunoelectrodiffusion |

## 群

| | | |
|---|---|---|
| 遺伝子群 ★ | イデンシグン | gene cluster |
| 吸収不良症候群 | キョウシュウフリョウ ショウコウグン | malabsorption syndrome |
| 空間群 | クウカングン | space group |
| 群体 | グンタイ | colony |
| 原発性免疫不全症候群 | ゲンハツセイメンエキフゼン ショウコウグン | primary immunodeficiency syndrome |
| 後天性免疫不全症候群 | コウテンセイメンエキフゼン ショウコウグン | acquired immune deficiency syndrome (AIDS) |
| 細胞性免疫不全症候群 | サイボウセイメンエキフゼン ショウコウグン | cell-mediated immunity deficiency syndrome |
| 実験群 | ジッケングン | experimental group |
| 症候群 ★ | ショウコウグン | syndrome |
| 増殖性分裂細胞群 | ゾウショクセイブンレツ サイボウグン | vegetative intermitotic cells |
| 対照群 ★ | タイショウグン | control group |
| 脱繊維素症候群 | ダツセンイソショウコウグン | defibrination syndrome |
| 長寿症候群 | チョウジュショウコウグン | longevity syndrome |
| 適応不全症候群 | テキオウフゼンショウコウグン | maladaptation syndrome |
| 不全症候群 | フゼンショウコウグン | deficiency syndrome |
| 分化性分裂細胞群 | ブンカセイブンレツサイボウグン | differentiating intermitotics |
| 分類群 | ブンルイグン | taxon |
| 免疫不全症候群 | メンエキフゼンショウコウグン | immunodeficiency syndrome |
| 連鎖群 | レンサグン | linkage group |

## 欠

| | | |
|---|---|---|
| 胃液欠乏症 | イエキケツボウショウ | achylia gastrica |
| 欠陥 ★ | ケッカン | defect |
| 欠陥ウイルス | ケッカンウイルス | defective virus |
| 欠陥干渉粒子 | ケッカンカンショウリュウシ | defective interfering particle |
| 欠陥ゲノム | ケッカンゲノム | defective genome |
| 欠陥ファージ | ケッカンファージ | defective phage |
| 欠陥溶原菌 | ケッカンヨウゲンキン | defective lysogen |
| 欠失 ★ | ケッシツ | deletion |
| 欠損 | ケッソン | defect |
| 欠損症 ★ | ケッソンショウ | deficiency |

| | | |
|---|---|---|
| 欠乏症 | ケツボウショウ | deficiency |
| 欠落 | ケツラク | deletion |
| 酵素欠損症 | コウソケッソンショウ | enzyme deficiency |
| 脱分枝酵素欠損症 | ダツブンシコウソケッソンショウ | debranching enzyme deficiency |
| 鉄欠乏性貧血 | テツケツボウセイヒンケツ | iron deficiency anemia |
| 分枝酵素欠損症 | ブンシコウソケッソンショウ | branching enzyme deficiency |
| 補欠分子族 | ホケツブンシゾク | prosthetic group |

## 座

| | | |
|---|---|---|
| 遺伝子座 ★ | イデンシザ | gene locus |
| 染色体間転座 | センショクタイカンテンザ | interchromosomal translocation |
| 染色体内転座 | センショクタイナイテンザ | intrachromosomal translocation |
| 相互転座 | ソウゴテンザ | reciprocal translocation |
| 転座 ★ | テンザ | translocation |
| 二座の | ニザの | bidentate |
| 捻れ舟型配座 | ねじれふながたハイザ | twisted boat conformation |
| 配座 ★ | ハイザ | conformation |
| 配座異性体 | ハイザイセイタイ | conformational isomer |
| 舟型配座 | ふながたハイザ | boat conformation |

## 始

| | | |
|---|---|---|
| 開始因子 | カイシインシ | initiation factor |
| 開始コドン ★ | カイシコドン | initiation codon |
| 開始tRNA | カイシtRNA | initiator tRNA |
| 開始複合体 | カイシフクゴウタイ | initiation complex |
| 鎖開始因子 | サカイシインシ | chain initiation factor |
| 始原型 ★ | シゲンがた | prototype |
| 始原生殖細胞 | シゲンセイショクサイボウ | primordial germ cell |
| 始原細胞 ★ | シゲンサイボウ | progenitor cell |
| 始原的生分子 | シゲンテキセイブンシ | primordial biomolecule |
| 創始者原理 | ソウシシャゲンリ | founder principle |
| 複製開始点 | フクセイカイシテン | origin of replication |

## 写

| | | |
|---|---|---|
| 逆転写 | ギャクテンシャ | reverse transcription |
| 逆転写酵素 | ギャクテンシャコウソ | reverse transcriptase |
| 抗転写終結因子 | コウテンシャシュウケツインシ | antitermination factor |
| 選択模写説 | センタクモシャセツ | copy choice hypothesis |
| 転写 ★ | テンシャ | transcription |

| | | |
|---|---|---|
| 転写一次産物 | テンシャイチジサンブツ | primary transcript |
| 転写因子 | テンシャインシ | transcription factor |
| 転写減衰 | テンシャゲンスイ | attenuation |
| 転写減衰因子 | テンシャゲンスイインシ | pausing factor |
| 転写酵素 | テンシャコウソ | transcriptase |
| 転写後修飾 | テンシャゴシュウショク | posttranscriptional modification |
| 転写後調節 | テンシャゴチョウセツ | posttranscriptional control |
| 転写後プロセシング | テンシャゴプロセシング | posttranscriptional processing |
| 転写終結 ★ | テンシャシュウケツ | transcription termination |
| 転写終結因子 | テンシャシュウケツインシ | transcription termination factor |
| 転写終結区 | テンシャシュウケツク | terminator |
| 転写調節 ★ | テンシャチョウセツ | transcriptional control |
| 転写調節因子 | テンシャチョウセツインシ | transcriptional control factor |

# 症

| | | |
|---|---|---|
| 過敏症 | カビンショウ | hypersensitivity |
| 感染症 ★ | カンセンショウ | infection, infectious disease |
| 欠損症 ★ | ケッソンショウ | deficiency |
| 原発性免疫不全症候群 | ゲンハツセイメンエキフゼンショウコウグン | primary immuno deficiency syndrome |
| 酵素欠損症 | コウソケッソンショウ | enzyme deficiency |
| 後天性免疫不全症候群 | コウテンセイメンエキフゼンショウコウグン | acquired immune deficiency syndrome (AIDS) |
| 細胞性免疫不全症候群 | サイボウセイメンエキフゼンショウコウグン | cell-mediated immunity deficiency syndrome |
| 症候群 ★ | ショウコウグン | syndrome |
| 脱分枝酵素欠損症 | ダツブンシコウソケッソンショウ | debranching enzyme deficiency |
| 適応不全症候群 | テキオウフゼンショウコウグン | maladaptation syndrome |
| 白血球減少症 | ハッケッキュウゲンショウショウ | leukopenia |
| 白血球増多症 | ハッケッキュウゾウタショウ | leukocytosis |
| 不全症候群 | フゼンショウコウグン | deficiency syndrome |
| 分枝酵素欠損症 | ブンシコウソケッソンショウ | branching enzyme deficiency |
| 免疫不全症 | メンエキフゼンショウ | immune disorder |
| 免疫不全症候群 | メンエキフゼンショウコウグン | immunodeficiency syndrome |

# 答

| | | |
|---|---|---|
| 一次応答 ★ | イチジオウトウ | primary response |
| 応答 ★ | オウトウ | response |
| 応答能 | オウトウノウ | competence |
| 緩和応答 | カンワオウトウ | relaxed response |
| 青色光応答 | セイショクコウオウトウ | blue light response |
| 二次応答 | ニジオウトウ | secondary response |
| 免疫応答 ★ | メンエキオウトウ | immune response |
| 免疫応答遺伝子 | メンエキオウトウイデンシ | immune response gene |
| 免疫学的不応答 | メンエキガクテキフオウトウ | immunological unresponsiveness |

# 頻

| | | |
|---|---|---|
| 遺伝子頻度 | イデンシヒンド | gene frequency |
| 組換え頻度 ★ | くみかえヒンド | recombination frequency |
| 高頻度形質導入溶菌液 | コウヒンドケイシツドウニュウヨウキンエキ | high-frequency transducting lysate |
| 対立遺伝子頻度 | タイリツイデンシヒンド | allele frequency |
| 低頻度形質導入 | テイヒンドケイシツドウニュウ | low-frequency transduction |
| 低頻度形質導入溶菌液 | テイヒンドケイシツドウニュウヨウキンエキ | low-frequency transducting lysate |
| 突然変異頻度 ★ | トツゼンヘンイヒンド | mutation frequency |
| 頻度 ★ | ヒンド | frequency |
| 頻度因子 | ヒンドインシ | frequency factor |
| 隣接塩基頻度分析 | リンセツエンキヒンドブンセキ | nearest-neighbor base frequency analysis |

*SUPPLEMENTARY VOCABULARY USING KANJI FROM CHAPTER 17*

| | | |
|---|---|---|
| 鋳型 | いがた | template |
| 異型接合体 | イケイセツゴウタイ | heterozygote |
| 異型染色体 | イケイセンショクタイ | heterosome |
| 遺伝子型 | イデンシがた | genotype |
| 遺伝子記号 | イデンシキゴウ | genetic symbol |
| 遺伝子組換え法 | イデンシくみかえホウ | gene recombination |
| 遺伝子置換 | イデンシチカン | gene substitution |
| 遺伝子変換 | イデンシヘンカン | gene conversion |
| 遺伝子量効果 | イデンシリョウコウカ | gene dosage effect |
| 遺伝的多型 | イデンテキタケイ | genetic polymorphism |
| 塩基置換 | エンキチカン | base substitution |
| 化学的突然変異 | カガクテキトツゼンヘンイ | chemical mutagenesis |

| | | |
|---|---|---|
| 誘発 | ユウハツ | |
| 核内低分子RNA | カクナイテイブンシRNA | small nuclear RNA |
| 記憶 | キオク | memory |
| 記号 | キゴウ | symbol |
| 機能細胞 | キノウサイボウ | functional cell |
| 組換え遺伝子 | くみかえイデンシ | recombination gene |
| 組換え体 | くみかえタイ | recombinant |
| 組換えDNA | くみかえDNA | recombinant DNA |
| 形質転換 | ケイシツテンカン | transformation |
| 減数分裂 | ゲンスウブンレツ | meiosis |
| 最適 | サイテキ | optimum |
| 自然形質転換 | シゼンケイシツテンカン | spontaneous transformation |
| 生物検定 | セイブツケンテイ | bioassay |
| 前進突然変異 | ゼンシントツゼンヘンイ | forward mutation |
| 相互的組換え | ソウゴテキくみかえ | reciprocal recombination |
| 相互乗換え | ソウゴのりかえ | reciprocal crossing over |
| 組織適合性 | ソシキテキゴウセイ | histocompatibility |
| 組織適合性遺伝子 | ソシキテキゴウセイイデンシ | histocompatibility gene |
| 多型性 | タケイセイ | polymorphism |
| 多分化能性幹細胞 | タブンカノウセイカンサイボウ | pluripotent(ial) stem cell |
| 通気装置 | ツウキソウチ | aerator |
| 低次形態 | テイジケイタイ | hypomorph |
| 適応 | テキオウ | adaptation |
| 適応酵素 | テキオウコウソ | adaptive enzyme |
| 転位 | テンイ | rearrangement, transition, translocation, transposition |
| 転移 | テンイ | metastasis, transfer, transition |
| 転移RNA | テンイRNA | transfer RNA |
| 転移酵素 | テンイコウソ | transferase |
| 転化 | テンカ | inversion |
| 転換 | テンカン | conversion, transversion |
| 点突然変異 | テントツゼンヘンイ | point mutation |
| 同型接合体 | ドウケイセツゴウタイ | homozygote |
| 糖転移酵素 | トウテンイコウソ | glycosyltransferase |
| 突然変異 | トツゼンヘンイ | mutation |
| 突然変異原性 | トツゼンヘンイゲンセイ | mutagenic, mutageneity |
| 突然変異固定 | トツゼンヘンイコテイ | mutation fixation |
| 突然変異体 | トツゼンヘンイタイ | mutant |
| 突然変異点 | トツゼンヘンイテン | mutation site |
| 突然変異誘発 | トツゼンヘンイユウハツ | mutagenesis |

| | | |
|---|---|---|
| 突然変異誘発物質 | トツゼンヘンイユウハツブッシツ | mutagen |
| 突然変異率 | トツゼンヘンイリツ | mutation rate |
| 熱安定酵素 | ネツアンテイコウソ | thermostable enyme |
| 乗換え | のりかえ | crossing over |
| 配列決定装置 | ハイレツケッテイソウチ | sequenator |
| 半減期 | ハンゲンキ | half-life |
| 繁殖不能性 | ハンショクフノウセイ | sterility, sterile |
| 微生物学的検定 | ビセイブツガクテキケンテイ | microbioassay |
| 表現型 | ヒョウゲンがた | phenotype |
| 表現型分散 | ヒョウゲンがたブンサン | phenotype variance |
| 部位特異的組換え | ブイトクイテキくみかえ | site-specific recombination |
| 部位特異的突然変異 | ブイトクイテキトツゼンヘンイ | site-specific mutation |
| 部位特異的突然変異誘発 | ブイトクイテキトツゼンヘンイユウハツ | site-directed mutagenesis |
| 不適合性 | フテキゴウセイ | incompatibility |
| 不適正塩基対 | フテキセイエンキツイ | mismatched base pair |
| 不等乗換え | フトウのりかえ | unequal crossing over |
| 普遍的組換え | フヘンテキくみかえ | general recombination |
| 分解能 | ブンカイノウ | resolution |
| 分化転換 | ブンカテンカン | transdifferentiation |
| 分布 | ブンプ | distribution |
| 変換 | ヘンカン | transmutation |
| 放射線突然変異生成 | ホウシャセントツゼンヘンイセイセイ | radiation mutagenesis |
| 放射能 | ホウシャノウ | radioactivity |
| 野生型 | ヤセイがた | wild type |
| 有糸分裂組換え | ユウシブンレツくみかえ | mitotic recombination |
| 有糸分裂乗換え | ユウシブンレツのりかえ | mitotic crossing over |
| 溶原変換 | ヨウゲンヘンカン | lysogenic conversion |
| 硫安 | リュウアン | ammonium sulfate |
| リン酸基転移酵素 | リンサンキテンイコウソ | phosphotransferase |

# EXERCISES

## Ex. 7.1 Matching Japanese and English terms

( ) 遺伝子座
( ) 遺伝子頻度
( ) 拡散因子
( ) 画線培養
( ) 組換え頻度
( ) 欠陥溶原菌

( ) 鎖開始因子
( ) 染色体間転座
( ) 促進拡散
( ) 対照群
( ) 適応不全症候群
( ) 転写酵素

( ) 転写終結因子
( ) 分画遠心分離
( ) 分枝酵素欠損症
( ) 免疫応答遺伝子
( ) 免疫不全症候群

1. branching enzyme deficiency
2. chain initiation factor
3. control group
4. defective lysogen
5. differential centrifugation
6. facilitated diffusion
7. gene frequency
8. immune response gene
9. immunodeficiency syndrome
10. interchromosomal translocation
11. locus
12. maladaptation syndrome
13. recombination frequency
14. spreading factor
15. streak culture
16. transcriptase
17. transcription termination factor

## Ex. 7.2 KANJI with the same ON reading

Look carefully at each of the two KANJI on the left, and note the ON reading that is common to both. Combine each KANJI on the left with the appropriate KANJI on the right to make a meaningful JUKUGO. Each technical term that contains one or more of the 100 KANJI introduced in this book can be found in the vocabulary lists for those KANJI. Other terms can be found in one of the supplementary vocabulary lists, including Lesson 0.

| | (1) | (2) | | |
|---|---|---|---|---|
| 1. | (1) 画 | (2) 拡 | 細胞分( )法 | 免疫( )散法 |
| 2. | (1) 緩 | (2) 寒 | ( )衝液 | ( )天培地 |
| 3. | (1) 欠 | (2) 血 | 酵素( )損症 | 白( )病細胞 |
| 4. | (1) 鎖 | (2) 差 | ( )延長因子 | 交( )画線培養 |
| 5. | (1) 写 | (2) 謝 | 基礎代( ) | 転( )後調節 |
| 6. | (1) 謝 | (2) 者 | 創始( )原理 | 代( )生成物 |
| 7. | (1) 終 | (2) 集 | ( )結因子 | 血球凝( ) |
| 8. | (1) 親 | (2) 神 | ( )経毒 | ( )和力 |
| 9. | (1) 走 | (2) 相 | ( )互転座 | ( )査 |
| 10. | (1) 答 | (2) 糖 | 一次免疫応( ) | ( )尿病 |
| 11. | (1) 遊 | (2) 誘 | 突然変異( )発物質 | 白血球( )走促進因子 |
| 12. | (1) 遊 | (2) 優 | 完全( )性 | ( )離細胞 |

### Ex. 7.3 Matching Japanese technical terms with definitions

Read each definition carefully, and then choose the appropriate technical term. Words that you have not yet encountered are listed following the definitions.

| | | |
|---|---|---|
| (　) 遺伝子頻度 | (　) 欠陥ファージ | (　) 免疫応答 |
| (　) 開始コドン | (　) 転座 | (　) 免疫拡散法 |
| (　) 画線培養 | (　) 転写 | (　) 連鎖群 |
| (　) 感染症 | | |

1. 固形培地の表面に菌を線状に接種して行う培養法。
2. 寒天などの支持体を用いて抗原と抗体を拡散させ、両者の濃度比が最適な条件になった位置に沈降線をつくらせる抗原抗体反応。
3. 染色体上に連鎖した一群の遺伝子のこと。
4. 遺伝子に欠失、置換、または点変異をもったり、形態形成が不完全のため、単独では宿主細胞に感染しても増殖できないファージ。
5. 染色体の一部が同じ染色体の他の部分に位置を変えたり、他の染色体上に位置を変える染色体異常現象。
6. 生体内での蛋白質合成に際して、mRNAの塩基配列の中で第1番目のアミノ酸を指定するコドン。
7. DNA依存性RNA合成が行われる過程。
8. 微生物、ウイルスにより起こされる病気。
9. 免疫によって生体内に抗体が生成すること。
10. ある遺伝子座で各種の対立遺伝子が一つのメンデル集団中に存在する相対的割合。

| | | |
|---|---|---|
| 支持体 | シジタイ | support |
| 濃度 | ノウド | concentration |
| 不完全 | フカンゼン | incompleteness |
| 単独 | タンドク | independence |
| 宿主 | シュクシュ | host |

**Ex. 7.4 Sentence translations**

Read each sentence carefully, and then translate it. Words that you have not yet encountered are listed following the sentences.

1. 各種の細菌の培養液をファージに対して交差画線培養を行うと、ファージ感受性菌では交差した部分が溶菌するが、非感受性菌ではファージに影響されずに増殖する。
2. 促進拡散では、まず基質Aと担体Cが生体膜の一方の表面で複合体C・Aを形成し、C・Aの形で膜を横切り、他の表面でAを解離してCはまた元に戻る。
3. 免疫応答に関するT細胞、B細胞、マクロファージ、抗体、補体などの減少・欠損あるいは機能異常があると、免疫応答能が低下するが、これには多くの原因があるので症候群としてまとめられ免疫不全症候群と呼ばれている。
4. 動物ウイルスを希釈しないで連続継代培養すると、自己増殖能を欠き、かつ元の完全ウイルスの増殖を阻害する欠陥ウイルスが出現する場合が多い。インフルエンザウイルスの欠陥干渉粒子は短い欠陥ゲノムを有する。
5. 一つの染色体上のある遺伝子座には一つの遺伝子が存在するが、それがとりうる状態は必ずしも一つとは限らない。そのため生物集団から多数の個体の抽出して特定の遺伝子座の状態を調べると、異なった遺伝子が見いだされることがある。
6. 原核細胞では開始tRNAはメチオニンを受容したあと、ホルミル化され、ホルミルメチオニルtRNA(fMet-tRNA$^{fMet}$)として蛋白質合成に用いられるが、真核細胞ではホルミル化酵素が存在しないので、Met-tRNA$^{Met}$のまま開始反応に用いられる。
7. 遺伝子がRNAポリメラーゼによって転写されて生じた転写一次産物(rRNA,tRNA,mRNAなどの前駆体)は種々の転写後プロセシングを受けて成熟RNAに変換される。
8. 最近、ある種の癌遺伝子産物が転写調節因子として機能していることが示されている。これに対して、原核細胞における転写調節因子としては、リプレッサーなどが知られ、DNAとの結合ならびに作用機作が調べられている。
9. I-AとI-E/C領域にそれぞれ位置する2個の遺伝子の相補作用による免疫応答の制御が観察されており、I-AとI-E/Cの2遺伝子支配によるハイブリッドIa抗原の存在がこの考えを支持している。
10. λファージの紫外線誘発溶菌液は、プロファージ座近傍の*gal*遺伝子などを$10^{-6}$という低頻度で形質導入するため、低頻度形質導入溶菌液と呼ばれる。

| | | |
|---|---|---|
| 交差 | コウサ | cross, crossing |
| 担体 | タンタイ | carrier |
| 複合体 | フクゴウタイ | complex |
| 横切る | よこぎる | to cross over, traverse |
| 解離 | カイリ | dissociation |
| まとめる | | to put together |
| 希釈 | キシャク | dilution |
| 自己増殖能 | ジコゾウショクノウ | ability to reproduce oneself |
| 抽出 | チュウシュツ | extraction |
| 受容 | ジュヨウ | receipt, acceptance |
| 癌遺伝子 | ガンイデンシ | oncogene |
| 作用機構 | サヨウキコウ | mechanism |
| 支持 | シジ | support |
| 近傍 | キンボウ | vicinity |

#### Ex. 7.5 Additional dictionary entries

( ) 遺伝子型　( ) 転移RNA　( ) 表現型
( ) 組換えDNA　( ) 突然変異誘発物質　( ) 分化転換
( ) 形質転換　( ) 乗換え　( ) 野生型
( ) 組織適合性

1. 与えられた環境の下で遺伝子の働きにより実際に現れる生物の形質。
2. 移植に際して拒絶反応が引き起こされるか起こされないかによって判定される宿主と移植片の間の遺伝的適合性。
3. 減数分裂で起こる相同染色体間の部分的交換で、連鎖している遺伝子の組換えを起こさせる現象。
4. 酵素などを用いて試験管内で異種のDNAを結合させることにより生成されたハイブリッド分子。
5. 生物の遺伝子や染色体における突然変異率を、自然突然変異が起こる率よりも高める効力をもつ物理的化学的外因子。
6. 生物の遺伝的基礎をなす遺伝子構成を指し、その特性を遺伝的に決定するもの。
7. 他の菌株から分離されたDNAによって、細菌の菌株の性質に変更を施す遺伝的組換え。
8. DNAを鋳型にして合成される分子で、蛋白質合成の際、アミノ酸の運搬を行う特徴をもつもの。
9. 特定の分化をすでに行っていた細胞や組織が質的に違った別のものに分化すること。
10. 野生(自然)集団中で最も高頻度に見られる型の系統、生物、遺伝子であり、多くの場合、突然変異型に対して優性のもの。

| | | | | | |
|---|---|---|---|---|---|
| 判定 | ハンテイ | judgement | 施す | ほどこす | to perform |
| 適合性 | テキゴウセイ | compatibility | 運搬 | ウンパン | conveyance, transport |
| 変更 | ヘンコウ | modification, revision | | | |

#### Ex. 7.6 Additional sentence translations

1. ある溶液が酸またはアルカリの添加によるpHの変化を和らげる作用をもつ時、つまりその際のpHの変化が純水の場合より小さい時、この作用を緩衝作用という。一般に弱酸とその塩、あるいは弱塩基とその塩の混合溶液は強い緩衝作用をもつので、緩衝液として、生化学や生物学の実験でのpHコントロールの目的で広く用いられている。生体物質の多くは弱酸か弱塩基であり、その緩衝作用が生体内のpHを恒常的に維持しており、ひいてはこの恒常性により生命を維持するという非常に重要な役割を果たしている。

2. 二本鎖RNAウイルスのRNAは端から端まで完全に二本鎖状体である。リボース-リン酸基が相互に繰返すRNAの背骨の糖部分に結合している塩基が、両方の鎖から二重らせんの内側に突出し、グアニンとシトシン、アデニンとウラシルが組合わさって水素結合により塩基対をつくっている。この塩基対はRNAのらせん軸に対して直角でなく少し傾いており、塩基対の中心もらせん軸の中心と重ならずらせんの少し内側に寄っていて、DNAのA形に近い二本鎖構造である。

3. 多細胞生物の生体内で起こる複雑な生命現象は、厳密に制御された細胞間協同作用、あるいは細胞と環境因子との相互作用により遂行されている。ホルモンや神経伝達物質とそれらの受容体との結合、免疫応答における自己、非自己の識別、特定の機能をもった細胞への分化、組織形成など、細胞間あるいは細胞と細胞外因子との認識が、引き続いて起こるさまざまな生体反応の引き金となる。

4. 走査型トンネル顕微鏡の金属のプローブを試料の表面に1nm以下に近づけ、試料とプローブの間に電圧をかけると両者の間にトンネル電流が流れる。トンネル電流を一定に保つように、圧電素子でプローブを上下させながら走査し、試料の表面の凹凸を正確になぞって試料表面の三次元像を得る。導電性の支持体の上に置かれた脂質、蛋白質、DNA、ウイルス粒子などの生体試料の観察もなされ、特にDNAについては二本鎖DNAのらせん構造や一本鎖DNAの塩基の化学構造が分かるような高分解能の像が得られている。

5. 動物のもつ免疫応答能を免疫寛容原、免疫抑制剤などで抑制することができる。この免疫抑制には抗原特異的なものと非特異的なものがあり、前者は免疫寛容を誘導できるような抗原によることが多く、後者はT、B細胞のような免疫を担当する細胞の分化分裂を阻害する免疫抑制剤によることが多い。また、抗原を投与する時免疫抑制剤を適用に組合せることにより特異的免疫抑制が成り立つ場合もある。

6. 細胞を破壊して細胞小器官を比較的純粋に分離する方法を細胞分画法という。多くの場合、冷温下で組織または細胞を機械的に破壊したのち、遠心機により、それぞれの小器官の沈降速度の差から、核、ミトコンドリア、リソソーム、リボソームなどに分離できる。また、比重の異なる液を直線、または段階的勾配にして調整しておき、懸濁液を重層して遠心する沈降平衡法によれば、比重の差で分画できる。

7. ある特定の酵素活性が、先天的に著しく低下、あるいは欠損した状態を酵素欠損症という。多くはDNAの構造異常のために、ペプチド鎖に変化が起こり、酵素活性にも異常を起こすものと考えられる。先天性代謝異常の大部分は酵素欠損症である。酵素欠損の結果、細胞の機能に直接影響を与える生化学的異常が生ずる。第一に基質の蓄積であり、アミノ酸や単糖類などの低分子化合物の場合、細胞外に流出し、血液・脳脊髄液・尿などの細胞外液中の濃度が上昇する。

8. 大腸菌の転写終結因子の作用によるRNA合成の停止を解除してRNA合成を下流に伸長させる因子がある。この抗転写終結因子の中で、lファージのN蛋白質とQ蛋白質が特に有名である。大腸菌にlファージが感染するとファージ遺伝子上の初期の転写終結信号でRNA合成が停止するが、感染直後に合成されるN蛋白質の働きにより転写終結が解除され、初期転写が続行する。

9. 生体が初めて接する抗原に対する免疫応答は、2度目以降に接した時とは異なる反応を示すことから、両者を区別して一次および二次免疫応答という。二次免疫応答は、体液性免疫で生産される抗体のクラスはIgGが主体で、血中抗体価もその上昇が急速であり、また細胞性免疫ではリンパ球機能の著しい亢進を示す。一次免疫応答との主要な違いは免疫記憶の成立の有無である。

10. 低頻度形質導入溶菌液中に含まれる形質導入ファージが、野生型ファージの存在下などで増殖した場合、高頻度形質導入溶菌液を生ずる。*gal*遺伝子を高頻度形質導入するλdgの高頻度形質導入溶菌液は特に有名である。また、P1Cmやφ80trpなどの、単独で増殖可能な特殊形質導入ファージも得られており、こうしたファージ溶菌液も高頻度形質導入溶菌液とよばれる。

| | | |
|---|---|---|
| 和らげる | やわらげる | to moderate, to alleviate |
| 純水 | ジュンスイ | pure water |
| ひいては | | in its turn |
| 端 | はし | end |
| 背骨 | せぼね | spinal column, backbone |
| 内側 | うちがわ | inside, inner part |
| 突出 | トッシュツ | projection, protrusion |
| 傾く | かたむく | to incline, to slant |
| 寄る | よる | to approach, to draw near |
| 厳密に | ゲンミツに | strictly |
| 協同 | キョウドウ | cooperation |
| 遂行 | スイコウ | accomplishment |
| 神経伝達物質 | シンケイデンタツブッシツ | neurotransmitter |
| 引き続いて | ひきつづいて | continuously |
| 引き金 | ひきがね | trigger |
| 保つ | たもつ | to preserve, to maintain |
| 凹凸 | オウトツ | irregularity, roughness |
| なぞる | | to trace, to follow |
| 像 | ゾウ | image |
| 免疫寛容原 | メンエキカンヨウゲン | tolerogen |
| 担当 | タントウ | (in) charge |
| 純粋 | ジュンスイ | purity |
| 段階的 | ダンカイテキ | stepwise |
| 重層する | ジュウソウする | to separate into layers |
| 脳脊髄液 | ノウセキズイエキ | cerebrospinal fluid |
| 尿 | ニョウ | urine |
| 上昇 | ジョウショウ | rise, increase |
| 解除 | カイジョ | removal, cancellation |
| 伸長 | シンチョウ | elongation |
| 有名な | ユウメイな | famous |
| 初期 | ショキ | early stage |
| 続行 | ゾッコウ | continuation |
| 初めて | はじめて | for the first time |
| 主体 | シュタイ | core, main part |
| 抗体価 | コウタイカ | antibody titer |
| 急速 | キュウソク | rapidity, swiftness |
| 亢進 | コウシン | increase in activity |

**Translations for Ex. 7.4**

1. If we carry out a cross streak culture with a phage on several varieties of bacteria, the portions of the phage-sensitive bacteria that intersect [with the phage] will undergo bacteriolysis, but the insensitive bacteria will grow unaffected by the phage.
2. In facilitated diffusion, substrate A and carrier C first form a complex C · A on one surface of a biomembrane. The complex then traverses the membrane in the form C · A. At the other surface A is released and C returns to its original state.
3. If there is a reduction in number, a defect or a functional irregularity in the T cells, B cells, macrophage, antibodies or complements related to an immune response, the immunological competence decreases. Since there are many factors, the symptoms are collected together as a syndrome, called the immunodeficiency syndrome.
4. If we continually subculture an animal virus without dilution, there are many instances in which a defective virus appears, one that lacks that ability to reproduce itself, and further, inhibits the growth of the complete virus. The defective interfering particles of the influenza virus possess a short defective genome.
5. In a particular locus on one chromosome there exists one gene, but there is not necessarily only one state that the gene can adopt. For that reason when we study the state of a specific locus by picking out several individuals from a biological population, there are instances when a different gene is detected.
6. In a procaryotic cell after the initiator tRNA accepts methionine it is formylated, and in the form of formylmethionyl-tRNA (fMet-$tRNA^{fMet}$) it is used for protein synthesis. However, the formylation enzyme is not found in a eucaryotic cell. Thus, in a eucaryotic cell it is used for the initiation reaction just as it is, in the form Met-$tRNA^{Met}$.
7. The primary transcripts (precursors of rRNA, tRNA, mRNA, and so on) that arise from the transcription of a gene by RNA polymerase undergo various forms of posttranscriptional processing and are transformed into mature RNA.
8. Recently, it has been indicated that certain varieties of oncogene products function as transcriptional control factors. In contrast, in procaryotic cells repressors, among others, are known to be transcriptional control factors; their linkage to DNA and their mechanism of action are being investigated.
9. Control of an immune response by the complementary action of two genes located in the I-A and I-E/C regions, respectively, has been observed. The existence of a hybrid Ia antigen resulting from two-gene control (I-A and I-EC) supports this thought.
10. The ultraviolet-induced lysate of the λ phage is called a low-frequency transducting lysate, because genes such as the *gal* gene in the vicinity of the prophage locus are transduced with low frequencies of $10^{-6}$.

# LESSON 8 第八課 BTJ 18

| | | |
|---|---|---|
| 域 | イキ | region; limit; level |
| 獲 | カク | acquisition, gain |
| 寛 | カン | leniency; generosity |
| 己 | コ | self |
| 主 | シュ | principal, primary, main |
| | シュ(として) | primarily, mainly |
| | おも(な) | principal, primary, main |
| 宿 | シュク | dwelling, lodging |
| 切 | セツ | cutting |
| | き(る) | to cut, sever {xu-verb} |
| | き(れる) | to be cut (off) |
| 末 | マツ | end; dust |
| | すえ | end |
| 輸 | ユ | transport |
| 領 | リョウ | territory; possession |

域 獲 寛 己 主 宿 切 末 輸 領

# 域

| | | |
|---|---|---|
| 下流領域 | カリュウリョウイキ | downstream region |
| 宿主域 | シュクシュイキ | host-range |
| 宿主域変異 ★ | シュクシュイキヘンイ | host-range mutation |
| 小量域寛容 | ショウリョウイキカンヨウ | low-zone tolerance |
| 上流領域 | ジョウリョウリョウイキ | upstream region |
| 先端領域 | センタンリョウイキ | proximal region |
| 相同性領域 | ソウドウセイリョウイキ | homology region |
| 大量域寛容 | タイリョウイキカンヨウ | high-zone tolerance |
| 帯領域沈降法 ★ | タイリョウイキチンコウホウ | band sedimentation |
| 嚢内領域 | ノウナイリョウイキ | intracisternal space |
| 末端領域 | マッタンリョウイキ | distal region |
| 領域 ★ | リョウイキ | region, domain |

# 獲

| | | |
|---|---|---|
| 獲得 ★ | カクトク | acquisition |
| 獲得寛容 ★ | カクトクカンヨウ | acquired tolerance |
| 獲得抵抗性 | カクトクテイコウセイ | acquired resistance |
| 獲得免疫 ★ | カクトクメンエキ | acquired immunity |
| 抗原性獲得 | コウゲンセイカクトク | antigen gain |
| 自然獲得免疫 | シゼンカクトクメンエキ | naturally acquired immunity |
| 人工獲得免疫 | ジンコウカクトクメンエキ | artificially acquired immunity |

# 寛

| | | |
|---|---|---|
| 獲得寛容 ★ | カクトクカンヨウ | acquired tolerance |
| 寛容 ★ | カンヨウ | tolerance |
| 寛容原 | カンヨウゲン | tolerogen |
| 寛容誘導 | カンヨウユウドウ | tolerance induction |
| 自己寛容 ★ | ジコカンヨウ | self tolerance, natural tolerance |
| 小量域寛容 | ショウリョウイキカンヨウ | low-zone tolerance |
| 大量域寛容 | タイリョウイキカンヨウ | high-zone tolerance |
| 免疫寛容 | メンエキカンヨウ | immunological tolerance |
| 養子寛容 | ヨウシカンヨウ | adoptive tolerance |

# 己

| | | |
|---|---|---|
| 自己感作 | ジコカンサ | autosensitization |
| 自己寛容 ★ | ジコカンヨウ | self tolerance, natural tolerance |
| 自己抗原 | ジココウゲン | autoantigen |
| 自己抗体 | ジココウタイ | autoantibody |
| 自己集合 ★ | ジコシュウゴウ | self-assembly |
| 自己消化 | ジコショウカ | autolysis |
| 自己消光 | ジコショウコウ | self-quenching |
| 自己触媒 | ジコショクバイ | autocatalysis |
| 自己増殖機械 | ジコゾウショクキカイ | automaton |
| 自己分解 | ジコブンカイ | autolysis |
| 自己分泌 | ジコブンピ[ツ] | autocrine |
| 自己免疫 ★ | ジコメンエキ | autoimmunity |
| 自己免疫疾患 | ジコメンエキシッカン | autoimmune disease |
| 自己溶菌 | ジコヨウキン | autolysis |
| 自己溶菌酵素 | ジコヨウキンコウソ | autolysin |
| 非自己抗原 | ヒジココウゲン | non-autoantigen |
| 利己的遺伝子 | リコテキイデンシ | selfish gene |

# 主

| | | |
|---|---|---|
| 移植片対宿主反応 ★ | イショクヘンタイシュクシュハンノウ | graft versus host reaction |
| 狭宿主性 | キョウシュクシュセイ | stenogenous |
| 宿主 ★ | シュクシュ | host, recipient |
| 宿主域 | シュクシュイキ | host-range |
| 宿主域変異 ★ | シュクシュイキヘンイ | host-range mutation |
| 宿主依存性修飾 | シュクシュイゾンセイシュウショク | host-controlled modification |
| 宿主因子 | シュクシュインシ | host factor |
| 宿主細胞回復 | シュクシュサイボウカイフク | host cell reactivation |
| 宿主ベクター系 | シュクシュベクターケイ | host vector system |
| 宿主誘導修飾 | シュクシュユウドウシュウショク | host-induced modification |
| 主鎖 | シュサ | main chain |
| 主体 | シュタイ | core, main part |
| 主要 | シュヨウ | principal |
| 主要組織適合性遺伝子複合体 | シュヨウソシキテキゴウセイイデンシフクゴウタイ | major histocompatibility complex |
| 主要組織適合性抗原系 | シュヨウソシキテキゴウセイコウゲンケイ | major histocompatibility system |
| 主抑制体 | シュヨクセイタイ | aporepressor |

## 宿

| | | |
|---|---|---|
| 移植片対宿主反応 ★ | イショクヘンタイシュクシュハンノウ | graft versus host reaction |
| 狭宿主性 | キョウシュクシュセイ | stenogenous |
| 宿主 ★ | シュクシュ | host, recipient |
| 宿主域 | シュクシュイキ | host-range |
| 宿主域変異 | シュクシュイキヘンイ | host-range mutation |
| 宿主依存性修飾 | シュクシュイゾンセイシュウショク | host-controlled modification |
| 宿主因子 | シュクシュインシ | host factor |
| 宿主細胞回復 ★ | シュクシュサイボウカイフク | host cell reactivation |
| 宿主ベクター系 | シュクシュベクターケイ | host vector system |
| 宿主誘導修飾 | シュクシュユウドウシュウショク | host-induced modification |

## 切

| | | |
|---|---|---|
| 枝切り酵素 | えだきりコウソ | debranching enzyme |
| 横断切片 | オウダンセッペン | cross section |
| 切れ目 ★ | きれめ | nick |
| 区切り | くぎり | punctuation, place to stop |
| 仕切る | しきる | to partition |
| 重感染切断 | ジュウカンセンセツダン | superinfection breakdown |
| 制限酵素切断地図 | セイゲンコウソセツダンチズ | restriction enzyme cleavage map |
| 切断 ★ | セツダン | cleavage |
| 切断再結合モデル | セツダンサイケツゴウモデル | breakage-reunion model |
| 切断地図 | セツダンチズ | cleavage map |
| 切片 ★ | セッペン | section |
| 切片法 | セッペンホウ | microtomy |
| 微細切断法 | ビサイセツダンホウ | microdissection |

## 末

| | | |
|---|---|---|
| 陥凹末端 | カンオウマッタン | recessed end |
| 終末槽 | シュウマツソウ | terminal cisterna |
| 神経終末 ★ | シンケイシュウマツ | nerve ending, synatpic ending |
| 相補末端 | ソウホマッタン | cohesive end |
| 粘着末端 | ネンチャクマッタン | cohesive end |
| 付着末端 | フチャクマッタン | cohesive end |
| 粉末 | フンマツ | powder |
| 末梢神経系 | マッショウシンケイケイ | peripheral nervous system |
| 末梢リンパ系組織 | マッショウリンパケイソシキ | peripheral lymphoid tissue |
| 末端 ★ | マッタン | terminal |

| | | |
|---|---|---|
| 末端基定量法 | マッタンキテイリョウホウ | end-group analysis |
| 末端重複 ★ | マッタンジュウフク | terminal repetition |
| 末端小粒 | マッタンショウリュウ | telomere |
| 末端分析 | マッタンブンセキ | terminal analysis |
| 末端領域 | マッタンリョウイキ | distal region |

# 輸

| | | |
|---|---|---|
| 角膜輪 | カクマクユ | arcus corneae |
| 経細胞輸送 | ケイサイボウユソウ | transcellular transport |
| 経上皮輸送 | ケイジョウヒユソウ | transepithelial transport |
| 細胞内輸送 | サイボウナイユソウ | intracellular transport |
| 軸索内輸送 | ジクサクナイユソウ | axoplasmic transport |
| 受動輸送 ★ | ジュドウユソウ | passive transport |
| 促進輸送 ★ | ソクシンユソウ | facilitated transport |
| 第一次能動輸送 | ダイイチジノウドウユソウ | primary active transport |
| 対向輸送 | タイコウユソウ | antiport |
| 第二次能動輸送 | ダイニジノウドウユソウ | secondary active transport |
| 担体輸送 | タンタイユソウ | carrier-mediated transport |
| 単輸送 | タンユソウ | uniport |
| 仲介輸送 | チュウカイユソウ | mediated transport |
| 同質細胞輸送 | ドウシツサイボウユソウ | homocellular transport |
| 能動輸送 | ノウドウユソウ | active transport |
| 能動輸送体 | ノウドウユソウタイ | pump |
| 膜動輸送 | マクドウユソウ | cytosis |
| 輸送 ★ | ユソウ | transport |
| 輸送性 | ユソウセイ | translocating |
| 輸送体 | ユソウタイ | carrier, translocator, transporter |
| 輸送蛋白質 | ユソウタンパクシツ | transport protein |
| 陽イオン輸送 | ヨウイオンユソウ | cation transport |
| リン酸輸送体 | リンサンユソウタイ | phosphate carrier |

# 領

| | | |
|---|---|---|
| 下流領域 ★ | カリュウリョウイキ | downstream region |
| 上流領域 ★ | ジョウリュウリョウイキ | upstream region |
| 先端領域 | センタンリョウイキ | proximal region |
| 相同性領域 | ソウドウセイリョウイキ | homology region |
| 帯領域沈降法 | タイリョウイキチンコウホウ | band sedimentation |
| 嚢内領域 | ノウナイリョウイキ | intracisternal space |
| 末端領域 | マッタンリョウイキ | distal region |
| 領域 ★ | リョウイキ | region, domain |

## *SUPPLEMENTARY VOCABULARY USING KANJI FROM CHAPTER 18*

| | | |
|---|---|---|
| アロ組織適合性抗原 | アロソシキテキゴウセイコウゲン | allo-histocompatibility antigen |
| 遺伝子重複 | イデンシジュウフク | gene duplication |
| 遺伝子複合体 | イデンシフクゴウタイ | gene complex |
| 遺伝地図 | イデンチズ | genetic map |
| 拮抗体 | キッコウタイ | antagonist |
| 近交系 | キンコウケイ | inbred strain |
| 抗原 | コウゲン | antigen |
| 抗原抗体反応 | コウゲンコウタイハンノウ | antigen-antibody reaction |
| 抗原抗体複合体 | コウゲンコウタイフクゴウタイ | antigen-antibody complex |
| 抗生物質 | コウセイブッシツ | antibiotic |
| 抗体 | コウタイ | antibody |
| 最大許容[線]量 | サイダイキョヨウ[セン]リョウ | maximum permissible dose |
| 細胞系 | サイボウケイ | cell line |
| 脂質二重層 | シシツニジュウソウ | lipid bilayer |
| 自然抗体 | シゼンコウタイ | natural antibody |
| 重複 | ジュウフク | duplication |
| 重複性 | ジュウフクセイ | redundancy |
| 受容体 | ジュヨウタイ | acceptor, receptor, recipient |
| 純系 | ジュンケイ | pure line |
| 神経伝達物質 | シンケイデンタツブッシツ | neurotransmitter |
| 制限酵素地図 | セイゲンコウソチズ | restriction map |
| 生殖系列 | セイショクケイレツ | germ cell line |
| 染色体地図 | センショクタイチズ | chromosome map |
| 臓器特異抗原 | ゾウキトクイコウゲン | organ-specific antigen |
| 組織定着抗体 | ソシキテイチャクコウタイ | tissue-bound antibody |
| 組織適合性抗原 | ソシキテキゴウセイコウゲン | histocompatibility antigen |
| 組織特異抗原 | ソシキトクイコウゲン | tissue-specific antigen |
| 対合複合体 | タイゴウフクゴウタイ | synaptonemal complex |
| 退縮 | タイシュク | regression |
| 多クローン性抗体 | タクローンセイコウタイ | polyclonal antibody |
| 多層増殖 | タソウゾウショク | multilayered growth |
| 単一特異性抗血清 | タンイツトクイセイコウケッセイ | monospecific antiserum |
| 単クローン性抗体 | タンクローンセイコウタイ | monoclonal antibody |
| 断層撮影法 | ダンソウサツエイホウ | tomography |
| 断面積 | ダンメンセキ | cross section |
| 地図 | チズ | map |
| 地図距離 | チズキョリ | map distance |
| 地図作成 | チズサクセイ | mapping |

地図単位　チズタンイ　map unit
抵抗性　テイコウセイ　resistance
点滴板　テンテキバン　spot plate
同一抗原　ドウイツコウゲン　homologous antigen
二重抗体法　ニジュウコウタイホウ　double antibody technique
乗換え地図　のりかえチズ　crossing over map
波長　ハチョウ　wavelength
光受容体　ひかりジュヨウタイ　photoreceptor
表面抗原　ヒョウメンコウゲン　surface antigen
複合体　フクゴウタイ　complex
複対立遺伝子　フクタイリツイデンシ　multiple alleles
物理的地図　ブツリテキチズ　physical map
部分抗原　ブブンコウゲン　hapten
分光光度計　ブンコウコウドケイ　spectro(photo)meter
平滑断端　ヘイカツダンタン　blunt end (of DNA)

## EXERCISES

### Ex. 8.1 Matching Japanese and English terms

| | | |
|---|---|---|
| ( ) 移植片対宿主反応 | ( ) 自然獲得免疫 | ( ) 大量域寛容 |
| ( ) 獲得寛容 | ( ) 宿主域変異 | ( ) 担体輸送 |
| ( ) 下流領域 | ( ) 主抑制体 | ( ) 末端基定量法 |
| ( ) 自己感作 | ( ) 制限酵素切断地図 | ( ) 末端領域 |
| ( ) 自己寛容 | ( ) 切片 | ( ) リン酸輸送体 |

| | | |
|---|---|---|
| 1. acquired tolerance | 6. downstream region | 11. naturally acquired immunity |
| 2. aporepressor | 7. end-group analysis | 12. phosphate carrier |
| 3. autosensitization | 8. graft versus host reaction | 13. restriction enzyme cleavage map |
| 4. carrier-mediated transport | 9. high-zone tolerance | 14. section |
| 5. distal region | 10. host-range mutation | 15. self tolerance |

### Ex. 8.2 KANJI with similar structural elements

Look carefully at each of the two KANJI on the left, and note which structural element is common to both. Combine each KANJI on the left with the appropriate KANJI on the right to make a meaningful JUKUGO. Each technical term that contains one or more of the 100 KANJI introduced in this book can be found in the vocabulary lists for those KANJI. Other terms can be found in one of the supplementary vocabulary lists, including Lesson 0.

| | | | |
|---|---|---|---|
| 1. (1) 域 (2) 培 | 栄養寒天( )養 | 小量( )寛容 |
| 2. (1) 感 (2) 減 | ( )覚性ニューロン | 転写( )衰 |
| 3. (1) 欠 (2) 次 | ( )陥ウイルス | 転写一( )産物 |
| 4. (1) 己 (2) 記 | 遺伝子( )号 | 自( )抗原 |
| 5. (1) 座 (2) 度 | 旋光( ) | 染色体内転( ) |
| 6. (1) 始 (2) 好 | 開( )コドン | ( )熱菌 |
| 7. (1) 写 (2) 与 | 供( )体 | 転( )調節 |
| 8. (1) 主 (2) 注 | 宿( )因子 | 皮下( )射 |
| 9. (1) 症 (2) 病 | 過敏( ) | ( )原体 |
| 10. (1) 答 (2) 合 | 遺伝子複( )体 | 免疫応( ) |
| 11. (1) 皮 (2) 波 | 上( ) | ( )長 |
| 12. (1) 補 (2) 複 | 遺伝子重( ) | ( )抑制物質 |
| 13. (1) 末 (2) 本 | 一( )鎖RNA | 粘着( )端 |
| 14. (1) 領 (2) 頻 | 相同性( )域 | 低( )度形質導入 |

**Ex. 8.3 Matching Japanese technical terms with definitions**

Read each definition carefully, and then choose the appropriate technical term. Words that you have not yet encountered are listed following the definitions.

( ) 獲得寛容
( ) 獲得免疫
( ) 自己抗体
( ) 宿主
( ) 宿主域
( ) 宿主誘導修飾
( ) 切断地図
( ) 帯領域沈降法
( ) 能動輸送
( ) 末端重複

1. 沈降係数の異なる溶質が別個の狭いゾーンを形成して沈降するような沈降速度法の一種。
2. ウイルスあるいはファージが感染し、増殖することができる細胞の種類。
3. 微生物感染、異物などの非自己抗原の刺激により後天的に獲得する特異的免疫。
4. 自然ないし先天的寛容に対して後天的に獲得された免疫寛容。
5. ある個体に生成された抗体で、その個体自身の構成物の抗原成分と反応する抗体。
6. 組換えDNA技術においてDNAが移入される生細胞、または寄生性生物の生育を許容する生物。
7. あるファージの増殖過程で、ファージ核酸が宿主菌から受ける非遺伝的な修飾反応のこと。
8. 制限酵素を利用してDNA上の各種制限酵素による切断位置を決定し、その分布と各遺伝子の位置を図示するというような物理的、化学的、ならびに酵素的手段によって作成される遺伝子地図。
9. ファージやウイルスの直線状ゲノムDNA(またはRNA(の右端と左端に同一の塩基配列が存在すること。
10. 生体膜を通して行われる物質の担体輸送の中で、膜内外での電気化学的ポテンシャルの勾配に逆行して輸送される現象。

| | | |
|---|---|---|
| 別個 | ベッコ | distinct, separate |
| 狭い | せまい | narrow |
| 刺激 | シゲキ | stimulus |
| 自身 | ジシン | itself, oneself |
| 移入 | イニュウ | introduction |
| 許容 | キョヨウ | permission, tolerance |
| 非遺伝的 | ヒイデンテキ | non-hereditary |
| 図示 | ズシ | illustration, graphical representation |
| 手段 | シュダン | means, measure |
| 右端 | みぎはし | right end |
| 左端 | ひだりはし | left end |
| 逆行 | ギャッコウ | reverse movement |

### Ex. 8.4 Sentence translations

Read each sentence carefully, and then translate it. Words that you have not yet encountered are listed following the sentences.

1. 小量域寛容ではTリンパ球のレベルで寛容が成立していて、Bリンパ球は寛容が成立していない。大量域寛容の場合に用いる抗原量との中間量では寛容状態が誘導されないのが特徴である。
2. ある種の病原体に対して、同種または抗原性共通で病原性の弱いものに感染しているため、免疫抗体・リンパ球ができて後天的に免疫を獲得することがある。
3. 人工的に特定の抗原を生体に接種し、抗体生産系を刺激し、接種された抗原物質(細菌、細菌類毒素、ウイルス)に対する抗体生成を起こすことによって成立する免疫現象を人工獲得免疫と呼ぶ。
4. 一般に同一個体内で自己の抗原に対する免疫応答は抑制されているが、自己免疫疾患では、異常なリンパ系組織の増殖が起こり、免疫応答も異常になり、自己抗原と反応する自己抗体が検出されるようになる。
5. 宿主とベクターの組合わせ、すなわち宿主ベクター系として、大腸菌K12株を宿主とするEK系、酵母菌を宿主とするSC系、枯草菌を宿主とするBS系さらに動植物培養細胞を宿主とする系がわが国の組換えDNA実験指針で認定されている。
6. ウイルスの受けた傷害が宿主細胞の修復系によって回復する現象を宿主細胞回復という。例えばファージに紫外線を照射し、菌に感染させるとファージDNAのピリミジン二量体は菌によって修復され、プラークを形成する場合がある。
7. クラスI制限酵素は染色体上の遺伝子でコードされる異なる3種のサブユニットから成り、ATP、$Mg^{2+}$、*S*-アデノシルメチオニン(AdoMet)を要求して、二本鎖DNAを非特異的に切断する。
8. DNAがある種の制限酵素で切断されたときに、数塩基の一本鎖部分をもつ末端が生じる。粘着末端どうしのホモロジーがDNA断端の連結に利用される。
9. 担体輸送のうちで、基質が膜内外での濃度勾配に従い、平衡に達するまで輸送されるものを促進拡散といい、濃度勾配に逆らって輸送されるものを能動輸送という。
10. 細胞内輸送は細胞質とは混じり合うことのない区画、すなわち細胞内顆粒の内腔を利用して行われ、しかもこの間種々のプロセシングが蛋白質やその糖鎖に加えられ、プレ蛋白質→成熟蛋白質といった変換が行われる。

| | | |
|---|---|---|
| 酵母菌 | コウボキン | *Saccharomyces cerevisiae* |
| 枯草菌 | コソウキン | *Bacillus subtilis* |
| わが国 | わがくに | Japan (N.B.: literally “our country”) |
| 指針 | シシン | guideline, protocol |
| 認定 | ニンテイ | recognition, approval |
| 断端 | ダンタン | fragment |
| 従う | したがう | to follow, to be in accordance with |
| 平衡 | ヘイコウ | equilibrium |
| 逆らう | さからう | to run counter to |
| 混じり合う | まじりあう | to be intermixed |
| 顆粒 | カリュウ | granule |
| 内腔 | ナイクウ | lumen |

**Ex. 8.5 Additional dictionary entries**

( ) 遺伝子重複
( ) 抗原
( ) 抗生物質
( ) 抗体
( ) 細胞系
( ) 脂質二重層
( ) 受容体
( ) 制限酵素地図
( ) 染色体地図
( ) 同一抗原

1. ある動物種の異なった個体の間で同じ組織、細胞あるいは生体抽出物が異なった抗原性を示す場合、これらの抗原物質を相対的に表現するために用いられる言葉。
2. 各種制限酵素を単独あるいは複数組合わせてDNAを切断し、得られたDNA断片をアガロースあるいはポリアクリルアミドゲル電気泳動法で分離後、各大きさを測定し、位置を決定することによって作られたもの。
3. 抗原刺激の結果、免疫反応によって生体内に生産される蛋白質で、免疫原(抗原)と特異的に結合する活性をもつもの。
4. 細胞に存在し、外来性の物質あるいは物理的刺激を認識して、細胞に応答を誘起する構造体。
5. 水溶液中で脂質、特にリン脂質のような極性脂質の親水性部分が直接水相に接し、疎水性部分が疎水結合によって互いに平行に並び、構造が二重になったもの。
6. 生物に侵入して抗体や感作リンパ球を生成させて体液性免疫や細胞性免疫を誘発する物質。
7. 染色体上に乗っている遺伝子の種類と配列順序や相互間の相対的距離関係を、一本の直線上に各遺伝子を一つの点で表して地図的に描いたもの。
8. 同一染色体の染色分体間の不等乗換え、または減数分裂期での相同染色体間の不等乗換えが起こり、一方の染色体で遺伝子の重複が生じて他方の染色体で欠失が生じるため、細胞が同じ遺伝子を二個以上もつ現象。
9. 微生物によって作られ、微生物の発育その他の機能を阻害する薬剤。
10. 広くは細胞における世代の連係、各種の進化の経路、種間の類縁関係のことで、培養細胞の場合、初代培養に存在した細胞または細胞群からの一連の系統のこと。

| | | |
|---|---|---|
| 抽出物 | チュウシュツブツ | extract(ive) |
| 言葉 | ことば | word |
| 並ぶ | ならぶ | to line up |
| 乗る | のる | to be recorded, to be carried |
| 距離 | キョリ | distance |
| 描く | えがく | to draw, to construct |
| 発育 | ハツイク | development, growth |
| 連係 | レンケイ | connection, linkage |
| 類縁 | ルイエン | affinity, family relation |

**Translations for Ex. 8.4**

1. In low-zone tolerance the tolerance arises at the T lymphocyte level but not for the B lymphocyte. One characteristic [of low-zone tolerance] is the fact that with a quantity of antigen that is intermediate between the levels used in low-zone and high-zone tolerance, a state of tolerance is not induced.
2. For certain types of pathogens there are occasions when immune antibodies or lymphocytes have been created due to [prior] infection by the same type of pathogen or a type that is weakly pathogenic and shares common antigenic characteristics [with the first type] so that [the individual] acquires immunity naturally.
3. Artificially acquired immunity is what we call the immune phenomenon that arises artificially as a result of the inoculation of an organism with a specific antibody and the [subsequent] stimulation of the antibody production system, [which in turn leads to] the production of antibodies against the antigenic substance (bacterium, bacterial toxin, virus) with which the organism was inoculated.
4. In general, an immune response within an individual toward its own antigens is suppressed. However, with autoimmune disease the growth of abnormal lymphatic tissue occurs, the immune response also becomes abnormal, and autoantibodies that react with autoantigens can now be detected.
5. The following systems are recognized in Japan's experiemental guidelines for recombinant DNA as combinations of host and vector, that is, host-vector systems: the EK system, which uses *Escherichia coli* strain K12 as host, the SC system, which uses *Saccharomyces cerevisiae* as host, the BS system, which uses *Bacillus subtilis* as host, as well as systems that use cultured plant or animal cells as host.
6. The phenomenon in which the injury sustained by a virus is repaired by the repair system of the host cell is called host cell reactivation. For example, there are instances in which if we irradiate a phage with ultraviolet light and then infect a bacterium, the pyrimidine dimer of the phage DNA is repaired by the bacterium, and plaque is formed.
7. Class I restriction enzymes are composed of three different types of subunits that are coded by genes on a chromosome. They seek out ATP, $Mg^{2+}$ and S-adenosylmethionine (AdoMet), and cleave double stranded DNA nonspecifically.
8. When DNA is cleaved by certain types of restriction enzymes, an end [of the DNA] that has a single-stranded portion [composed of] several bases is created. The homology between cohesive ends is utilized in connecting DNA fragments.
9. Among [the forms of] carrier-mediated transport the situation in which the substrate is transported until equilibrium is achieved in accordance with the concentration gradient outside and inside the membrane is called facilitated diffusion. The situation in which the substrate is transported counter to the concentration gradient is called active transport.
10. Intracellular transport is carried out utilizing a compartment that is not intermixed with cytoplasm, in other words the lumen of an intercellular granule. Further, during this time various types of processing are applied to a protein and its sugar chains, and the transformation from pre-protein to mature protein takes place.

| | | |
| --- | --- | --- |
| 炎 | エン | inflammation; flame |
| | ほのお | flame |
| 癌 | ガン | cancer |
| 競 | キョウ | competition |
| 殺 | サツ | killing |
| | ころ(す) | to kill |
| 死 | シ | death |
| | し(ぬ) | to die |
| 腫 | シュ | tumor, swelling |
| | は(らす) | to cause to swell, inflame |
| | は(れる) | to swell, become swollen |
| 耐 | タイ- | -resistant, -proof |
| | た(える) | to endure |
| 致 | チ | causing, attaining |
| | いた(す) | to do {humble} |
| 読 | ドク | reading |
| | よ(む) | to read |
| | よ(める) | to be legible |
| 瘍 | ヨウ | boil, carbuncle |

炎 癌

競 殺

死 腫

耐 致

読 瘍

## 炎

| | | |
|---|---|---|
| 炎光光度計 | エンコウコウドケイ | flame photometer |
| 炎光光度検出器 | エンコウコウドケンシュツキ | flame photometric detector |
| 炎光分光分析 | エンコウブンコウブンセキ | flame spectrochemical analysis |
| 炎症 ★ | エンショウ | inflammation |
| 肝炎 | カンエン | hepatitis |
| 関節炎 ★ | カンセツエン | arthritis |
| 起脳炎蛋白質 | キノウエンタンパクシツ | encephalitogenic protein |
| 結節性全脳炎 | ケッセツセイゼンノウエン | nodular panencephalitis |
| 限局性回腸炎 | ゲンキョクセイカイチョウエン | regional ileitis |
| 糸球体腎炎 | シキュウタイジンエン | glomerulonephritis |
| 消炎鎮痛剤 | ショウエンチンツウザイ | antiphlogistic sedative drug |
| 腎炎 | ジンエン | nephritis |
| 腎炎因子 | ジンエンインシ | nephritic factor |
| 水素炎イオン化検出器 | スイソエンイオンカケンシュツキ | flame ionization detector |
| 水疱性口内炎ウイルス | スイホウセイコウナイエンウイルス | vesicular stomatitis virus |
| 多発性神経炎 | タハツセイシンケイエン | polyneuropathy |
| 脳炎 | ノウエン | encephalitis |
| 肺炎 ★ | ハイエン | pneumonia |
| 肺炎双球菌 | ハイエンソウキュウキン | *Diplococcus pneumoniae* |
| 馬杉腎炎 | ますぎジンエン | Masugi's nephritis |
| 狼瘡性腎炎 | ロウソウセイジンエン | lupus nephritis |

## 癌

| | | |
|---|---|---|
| 化学発癌 | カガクハツガン | chemical carcinogenesis |
| 癌遺伝子 ★ | ガンイデンシ | oncogene |
| 癌ウイルス | ガンウイルス | oncogenic virus |
| 癌化 ★ | ガンカ | malignant transformation |
| 肝癌 | カンガン | hepatic carcinoma, hepatoma |
| 癌原遺伝子 | ガンゲンイデンシ | proto-oncogene |
| 癌腫 | ガンシュ | carcinoma |
| 癌性悪液質 | ガンセイアクエキシツ | cancerous cachexia |
| 癌胎児性抗原 | ガンタイジセイコウゲン | carcinoembryonic antigen |
| 癌様体 | ガンヨウタイ | carcinoid |
| 奇形癌 | キケイガン | teratocarcinoma |
| 原発性肝癌 | ゲンハッセイカンガン | primary hepatic carcinoma |
| 抗癌剤 | コウガンザイ | anticancer agent |
| 黒色癌前駆体 | コクショクガンゼンクタイ | lentigo maligna |

| | | |
|---|---|---|
| 最小偏奇肝癌 | サイショウヘンキカンガン | minimal deviation hepatoma |
| 上咽頭癌 | ジョウイントウガン | nasopharyngeal carcinoma |
| 助発癌物質 | ジョハツガンブッシツ | auxiliary carcinogen |
| 制癌物質 | セイガンブッシツ | cancerocidal substance |
| 前癌病変 | ゼンガンビョウヘン | precancerous change |
| 胆管癌 | タンカンガン | cholangioma |
| 単独発癌物質 | タンドクハツガンブッシツ | solitary carcinogen |
| 直接型発癌物質 | チョクセツがたハツガンブッシツ | direct carcinogen |
| 発癌 ★ | ハツガン | carcinogenesis |
| 発癌性 | ハツガンセイ | carcinogenic |
| 発癌促進物質 | ハツガンソクシンブッシツ | promoter of carcinogenesis |
| 発癌物質 | ハツガンブッシツ | carcinogen |
| 発癌プロモーター | ハツガンプロモーター | promoter of carcinogenesis |
| 不完全発癌物質 | フカンゼンハツガンブッシツ | incomplete carcinogen |
| 腹水癌 | フクスイガン | ascites carcinoma |
| 放射線発癌 | ホウシャセンハツガン | radiation carcinogenesis |

# 競

| | | |
|---|---|---|
| 競合 ★ | キョウゴウ | competition |
| 競合アッセイ | キョウゴウアッセイ | competitive assay |
| 競合阻害 ★ | キョウゴウソガイ | competitive inhibition |
| 競合阻害剤 | キョウゴウソガイざい | competitive inhibitor |
| 競合的拮抗薬 | キョウゴウテキキッコウヤク | competitive antagonist |
| 競争 ★ | キョウソウ | competition |
| 抗原競合 | コウゲンキョウゴウ | antigenic competition |
| 非競合阻害 | ヒキョウゴウソガイ | noncompetitive inhibition |
| 非競合的 | ヒキョウゴウテキ | noncompetitive |
| 非競合的拮抗薬 | ヒキョウゴウテキキッコウヤク | noncompetitive antagonist |
| 不競合阻害 | フキョウゴウソガイ | uncompetitive inhibition |
| 不競合的 | フキョウゴウテキ | uncompetitive |

# 殺

| | | |
|---|---|---|
| 殺菌 ★ | サッキン | killing bacteria, killing fungi, killing germs, sterilization |
| 殺菌剤 ★ | サッキンザイ | bactericide, fungicide, germicide, microbicide disinfectant |
| 殺菌性 | サッキンセイ | bactericidal |
| 殺菌薬 | サッキンヤク | germicide |

| | | |
|---|---|---|
| 殺線虫薬 | サツセンチュウヤク | nematicide |
| 殺草効果 | サッソウコウカ | herbicidal effect |
| 殺草作用 | サッソウサヨウ | herbicidal action |
| 殺虫剤 ★ | サッチュウザイ | insecticide |
| 殺虫殺鼠剤 | サッチュウサッソザイ | pesticide |
| 殺壁蝨剤 | サツダニザイ | acaricide |
| 自殺基質 | ジサツキシツ | suicide substrate |
| 低温殺菌 | テイオンサッキン | pasteurization |
| 屠殺 | トサツ | sacrifice, slaughter |
| 屠殺場 | トサツジョウ | slaughterhouse |
| 放射線殺菌 | ホウシャセンサッキン | radiation sterilization |

## 死

| | | |
|---|---|---|
| 亜致死性損傷 | アチシセイソンショウ | sublethal damage |
| 壊死 ★ | エシ | necrosis |
| 壊死過程 | エシカテイ | necrobiosis |
| 仮死 | カシ | asphyxia |
| 細胞生死判別試験 | サイボウセイシハンベツシケン | cell viability test |
| 死骸 | シガイ | cadaver |
| 死滅 ★ | シメツ | death, extinction |
| 腫瘍壊死因子 | シュヨウエシインシ | tumor necrosis factor |
| 条件致死変異 | ジョウケンチシヘンイ | conditional lethal mutation |
| 条件致死変異体 | ジョウケンチシヘンイタイ | conditional lethal mutant |
| 接合体致死 | セツゴウタイチシ | zygotic lethal |
| 致死遺伝子 ★ | チシイデンシ | lethal gene |
| 致死因子 | チシインシ | lethal factor |
| 致死合成 | チシゴウセイ | lethal synthesis |
| 致死雑種 | チシザッシュ | lethal hybrid |
| 致死線量 | チシセンリョウ | lethal dose |
| 致死相当量 | チシソウトウリョウ | lethal equivalent |
| 致死的 | チシテキ | lethal, fatal |
| 致死突然変異 | チシトツゼンヘンイ | lethal mutation |
| 半致死時間 | ハンチシジカン | fifty percent lethal time |
| 半致死線量 | ハンチシセンリョウ | fifty percent lethal dose |
| 不死化 | フシカ | immortalization |
| 崩壊死 | ホウカイシ | decay suicide |
| 優性致死遺伝子 | ユウセイチシイデンシ | dominant lethal gene |
| 劣性致死遺伝子 | レッセイチシイデンシ | recessive lethal gene |

# 腫

| | | |
|---|---|---|
| 悪性腫瘍 | アクセイシュヨウ | malignant tumor |
| 遺伝性血管神経性浮腫 | イデンセイケッカンシンケイセイフシュ | hereditary angioneurotic edema |
| 黄色腫 | オウショクシュ | xanthoma |
| 黄色腫症 | オウショクシュショウ | xanthomatosis |
| 褐色細胞腫 | カッショクサイボウシュ | pheochromocytoma |
| 癌腫 ★ | ガンシュ | carcinoma |
| 奇形腫 | キケイシュ | teratoma |
| 棘細胞腫 | キョクサイボウシュ | acanthoma |
| 形質細胞腫 | ケイシツサイボウシュ | plasmocytoma |
| 結節型悪性黒色腫 | ケッセツがたアクセイコクショクシュ | nodular malignant melanoma |
| 結節型黄色腫 | ケッセツがたオウショクシュ | xanthoma tuberosum |
| 抗腫瘍免疫 | コウシュヨウメンエキ | antitumor immunity |
| 黒色腫 | コクショクシュ | melanoma |
| 孤立性形質細胞腫 | コリツセイケイシツサイボウシュ | solitary plasmacytoma |
| 粥腫 | ジュクシュ | atheroma |
| 腫脹 | シュチョウ | swelling |
| 腫瘍 ★ | シュヨウ | tumor |
| 腫瘍壊死因子 | シュヨウエシインシ | tumor necrosis factor |
| 腫瘍化 | シュヨウカ | tumorigenic transformation |
| 腫瘍抗原 ★ | シュヨウコウゲン | tumor antigen |
| 腫瘍細胞 | シュヨウサイボウ | tumor cell |
| 腫瘍特異抗原 | シュヨウトクイコウゲン | tumor-specific antigen |
| 鳥肉腫ウイルス | とりニクシュウイルス | avian sarcoma virus |
| 肉腫 | ニクシュ | sarcoma |
| 肉腫遺伝子 | ニクシュイデンシ | sarcoma gene |
| 肉腫ウイルス | ニクシュウイルス | sarcoma virus |
| 乳頭腫 | ニュウトウシュ | papilloma |
| 乳頭腫ウイルス | ニュウトウシュウイルス | papilloma virus |
| 粘液水腫 | ネンエキスイシュ | myxedma |
| 肥満細胞腫 | ヒマンサイボウシュ | mastocytoma |
| 浮腫 | フシュ | edema |
| 扁平型黄色腫 | ヘンペイがたオウショクシュ | xanthoma planum |
| 無色性黒色腫 | ムショクセイコクショクシュ | amelanotic melanoma |
| 融合雑種腫瘍細胞 | ユウゴウザッシュシュヨウサイボウ | hybridoma |
| 卵巣性奇形腫 | ランソウセイキケイシュ | ovarian teratoma |
| 良性腫瘍 | リョウセイシュヨウ | benign tumor |
| 類肉腫症 | ルイニクシュショウ | sarcoidosis |

## 耐

| | | |
|---|---|---|
| 速成耐性 | ソクセイタイセイ | tachyphylaxis |
| 耐寒性 | タイカンセイ | cryotolerance |
| 耐性 ★ | タイセイ | resistance |
| 耐性細胞 | タイセイサイボウ | resistant cell |
| 耐熱性 ★ | タイネツセイ | thermostability |
| 多剤耐性 | タザイタイセイ | multiple drug resistance |
| 多剤耐性因子 | タザイタイセイインシ | multiple drug resistance factor |
| 不耐症 ★ | フタイショウ | intolerance |
| 薬剤耐性 | ヤクザイタイセイ | drug resistance |

## 致

| | | |
|---|---|---|
| 亜致死性損傷 | アチシセイソンショウ | sublethal damage |
| 条件致死変異 | ジョウケンチシヘンイ | conditional lethal mutation |
| 条件致死変異体 | ジョウケンチシヘンイタイ | conditional lethal mutant |
| 接合体致死 | セツゴウタイチシ | zygotic lethal |
| 致死遺伝子 ★ | チシイデンシ | lethal gene |
| 致死因子 | チシインシ | lethal factor |
| 致死合成 | チシゴウセイ | lethal synthesis |
| 致死雑種 | チシザッシュ | lethal hybrid |
| 致死線量 ★ | チシセンリョウ | lethal dose |
| 致死相当量 | チシソウトウリョウ | lethal equivalent |
| 致死的 | チシテキ | lethal, fatal |
| 致死突然変異 ★ | チシトツゼンヘンイ | lethal mutation |
| 半致死時間 | ハンチシジカン | fifty percent lethal time |
| 半致死線量 | ハンチシセンリョウ | fifty percent lethal dose |
| 優性致死遺伝子 | ユウセイチシイデンシ | dominant lethal gene |
| 劣性致死遺伝子 | レッセイチシイデンシ | recessive lethal gene |

## 読

| | | |
|---|---|---|
| 解読 | カイドク | decoding |
| 誤読 | ゴドク | reading mistake |
| 読み終わりコドン ★ | よみおわりコドン | chain-terminating codon |
| 読み過ごし転写 | よみすごしテンシャ | read-through transcription |
| 読み通し | よみとおし | read through |
| 読み取る | よみとる | read (and comprehend) |
| 読み始めコドン ★ | よみはじめコドン | chain-initiating codon |
| 読み枠 ★ | よみわく | reading frame |
| 読み枠突然変異 | よみわくトツゼンヘンイ | reading frame mutation |

## 瘍

| | | |
|---|---|---|
| 悪性腫瘍 ★ | アクセイシュヨウ | malignant tumor |
| 抗腫瘍免疫 | コウシュヨウメンエキ | antitumor immunity |
| 腫瘍 ★ | シュヨウ | tumor |
| 腫瘍壊死因子 | シュヨウエシインシ | tumor necrosis factor |
| 腫瘍化 | シュヨウカ | tumorigenic transformation |
| 腫瘍抗原 | シュヨウコウゲン | tumor antigen |
| 腫瘍細胞 | シュヨウサイボウ | tumor cell |
| 腫瘍特異抗原 | シュヨウトクイコウゲン | tumor-specific antigen |
| 消化性潰瘍 | ショウカセイカイヨウ | peptic ulcer |
| 融合雑種腫瘍細胞 | ユウゴウザッシュシュヨウサイボウ | hybridoma |
| 良性腫瘍 ★ | リョウセイシュヨウ | benign tumor |

### *SUPPLEMENTARY VOCABULARY USING KANJI FROM CHAPTER 19*

| | | |
|---|---|---|
| 核磁気共鳴 | カクジキキョウメイ | nuclear magnetic resonance |
| 活性 | カッセイ | activity |
| 活性汚泥 | カッセイオデイ | activated sludge |
| 活性汚泥法 | カッセイオデイホウ | activated sludge process |
| 活性中心 | カッセイチュウシン | active center |
| 逆位 | ギャクイ | inversion |
| 逆方向の遺伝学 | ギャクホウコウのイデンガク | reverse genetics |
| 互変異性酵素 | ゴヘンイセイコウソ | tautomerase |
| 混合物 | コンゴウブツ | mixture |
| 試験管内遺伝学 | シケンカンナイイデンガク | in vitro genetics |
| 試験管内組換え | シケンカンナイくみかえ | in vitro recombination |
| 磁気共鳴 | ジキキョウメイ | magnetic resonance |
| 精製 | セイセイ | purification |
| 製法 | セイホウ | preparation |
| 超音波処理 | チョウオンパショリ | ultrasonification |
| 超コイルDNA | チョウコイルDNA | supercoiled DNA |
| 二方向複製 | ニホウコウフクセイ | bidirectional replication |
| 濃度 | ノウド | concentration |
| 半保存的複製 | ハンホゾンテキフクセイ | semiconservative replication |
| 比活性 | ヒカッセイ | specific activity |
| 非共有結合 | ヒキョウユウケツゴウ | noncovalent bond |
| 非極性基 | ヒキョクセイキ | nonpolar group |
| 非極性溶媒 | ヒキョクセイヨウバイ | nonpolar solvent |

| | | |
|---|---|---|
| 非相互組換え | ヒソウゴくみかえ | nonreciprocal recombination |
| 非相同的組換え | ヒソウドウテキくみかえ | nonhomologous recombination |
| 非特異的な | ヒトクイテキな | nonspecific |
| 非反復配列 | ヒハンプクハイレツ | nonrepetitive sequence |
| 表現型混合 | ヒョウゲンがたコンゴウ | phenotypic mixing |
| 付加 | フカ | addition |
| 不可逆 | フカギャク | irreversible |
| 不活性化 | フカッセイカ | inactivation |
| 付加突然変異 | フカトツゼンヘンイ | addition mutation |
| 付加反応 | フカハンノウ | addition reaction |
| 付随体 | フズイタイ | satellite |
| 付随DNA | フズイDNA | satellite DNA |
| 付着 | フチャク | attachment |
| 複製 | フクセイ | duplication, replication |
| 複製型 | フクセイがた | replicative form |
| 複製子 | フクセイシ | replicon |
| 不連続複製 | フレンゾクフクセイ | discontinuous replication |
| 分子活性 | ブンシカッセイ | molecular activity |
| 分子間相互作用 | ブンシカンソウゴサヨウ | intermolecular interaction |
| 薬剤抵抗性 | ヤクザイテイコウセイ | drug resistance |
| 薬剤抵抗性遺伝子 | ヤクザイテイコウセイイデンシ | drug resistance gene |
| 薬剤抵抗性細胞 | ヤクザイテイコウセイサイボウ | drug resistant cell |
| 拮抗薬 | キッコウヤク | antagonist |

## EXERCISES

### Ex. 9.1 Matching Japanese and English terms

| | | |
|---|---|---|
| ( ) 悪性腫瘍 | ( ) 競合的拮抗薬 | ( ) 放射線殺菌 |
| ( ) 炎症 | ( ) 耐性細胞 | ( ) 薬剤耐性 |
| ( ) 解読 | ( ) 致死線量 | ( ) 優性致死遺伝子 |
| ( ) 癌遺伝子 | ( ) 低温殺菌 | ( ) 読み通し |
| ( ) 癌腫 | ( ) 肺炎 | ( ) 良性腫瘍 |
| ( ) 競合阻害 | ( ) 発癌促進物質 | |

1. benign tumor
2. carcinoma
3. competitive antagonist
4. competitive inhibition
5. decoding
6. dominant lethal gene
7. drug resistance
8. inflammation
9. lethal dose
10. malignant tumor
11. oncogene
12. pasteurization
13. pneumonia
14. promoter of carcinogenesis
15. radiation sterilization
16. read through
17. resistant cell

### Ex. 9.2 KANJI with the same ON reading

Look carefully at each of the two KANJI on the left, and note the ON reading that is common to both. Combine each KANJI on the left with the appropriate KANJI on the right to make a meaningful JUKUGO. Each technical term that contains one or more of the 100 KANJI introduced in this book can be found in the vocabulary lists for those KANJI. Other terms can be found in one of the supplementary vocabulary lists, including Lesson 0.

| | | | |
|---|---|---|---|
| 1. | (1) 炎 (2) 塩 | 関節( ) | ( )基対 |
| 2. | (1) 獲 (2) 拡 | ( )散係数 | ( )得抵抗性 |
| 3. | (1) 寛 (2) 寒 | 耐( )性 | 免疫( )容 |
| 4. | (1) 癌 (2) 含 | 発( )物質 | らせん( )量 |
| 5. | (1) 競 (2) 共 | 核磁気( )鳴 | 非( )合阻害 |
| 6. | (1) 己 (2) 呼 | ( )吸 | 自( )抗体 |
| 7. | (1) 好 (2) 抗 | ( )気性生物 | 組織特異( )原 |
| 8. | (1) 腫 (2) 種 | 致死雑( ) | 肉( )遺伝子 |
| 9. | (1) 宿 (2) 縮 | ( )主 | 退( ) |
| 10. | (1) 切 (2) 節 | ( )断地図 | 翻訳調( ) |
| 11. | (1) 耐 (2) 対 | ( )照群 | ( )熱性 |
| 12. | (1) 読 (2) 毒 | 誤( ) | ( )素 |
| 13. | (1) 瘍 (2) 養 | 凝固血漿培( ) | 腫( )細胞 |
| 14. | (1) 領 (2) 量 | 遺伝子( )効果 | 上流( )域 |

**Ex. 9.3 Matching Japanese technical terms with definitions**

Read each definition carefully, and then choose the appropriate technical term. Words that you have not yet encountered are listed following the definitions.

| | | |
|---|---|---|
| (　) 炎症 | (　) 耐性細胞 | (　) 非競合阻害 |
| (　) 癌遺伝子 | (　) 致死遺伝子 | (　) 放射線殺菌 |
| (　) 癌腫 | (　) 致死変異 | (　) 読み枠 |
| (　) 腫瘍細胞 | | |

1. 血液中の蛋白質や免疫担当細胞などが感染や損傷に対応して組織内に侵入してくるプロセス。
2. 細胞が癌化する際に、癌化の指令を発する遺伝子。
3. 悪性腫瘍のうち、その起源が上皮性細胞であるもの。
4. 阻害剤が遊離の酵素と酵素-基質複合体の両者に結合することによって生じる阻害様式のこと。
5. ガンマ線または電子線の照射による殺菌。
6. 個体が成長の途中の段階で死に至る突然変異。
7. 欠失または突然変異によって、個体の正常な発育を阻害し、個体の生殖年齢以前に死を起こす遺伝子。
8. 正常細胞が遺伝子レベルの変化によって安定した形質として異常増殖能を獲得したもの。
9. 致死作用をもつ要因に対して抵抗性をもつ細胞。
10. mRNA上の塩基の配列が遺伝情報として蛋白質に翻訳される際に読み取られていく区切り。

| | | | | | |
|---|---|---|---|---|---|
| 起源 | キゲン | origin | 至る | いたる | to lead to, to result in |
| 途中 | トチュウ | on the way, *en route* | 年齢 | ネンレイ | age |

**Ex. 9.4 Sentence translations**

Read each sentence carefully, and then translate it. Words that you have not yet encountered are listed following the sentences.

1. 炎症の経時的変化をみると、組織傷害直後に一過性の細動脈収縮が起こり、ついで細動脈拡張、毛細血管・細静脈の拡張がみられる。血管内から傷害組織中へ血漿成分の浸出が起こる。
2. 細胞染色体上の癌原遺伝子は、RNA腫瘍ウイルスのゲノムに挿入され、染色体上で変化して細胞癌化能を有するようになるが、一方で生物種を越えて普遍的に存在することが知られており、生命現象を営む上で基本的な役割を果たしている。
3. 発癌物質の多くは、発癌イニシエーターとしての作用と発癌プロモーターとしての作用を同時にもっているが、中にはプロモーター作用が弱く、助発癌物質の助けを借りないと発癌を行い得ないものもある。
4. 酵素活性は、種々の化合物によって触媒活性が可逆的に阻害される場合が多い。酵素活性の阻害にはいろいろな形式があるが、阻害剤分子が基質分子と競合して、基質結合部位を取り合う型のものが競合阻害と呼ばれている。

5. 病原微生物を死滅させるために用いる薬物が多い。器具などに付着した菌に対して用いる薬物を消毒薬、生きた組織に適用する薬物を殺菌薬、腐敗を防ぐための薬物を防腐薬と称するが、厳密な区別はない。
6. 劣性致死遺伝子はホモ接合体になって初めてその効果がでるが、優性致死遺伝子はその遺伝子をもつ個体をすべて殺す。
7. 良性腫瘍は、腫瘍細胞およびその配列がその由来する正常細胞に近い形態をとり、浸潤性・転移性のない腫瘍である。悪性腫瘍は、腫瘍細胞の形態や配列が種々の点で元の細胞と異なっており、転移性・浸潤性があって放置すれば必ず致死的である腫瘍である。
8. 腫瘍特異抗原は宿主にとって異物であるので、これに対する特異的免疫反応が起こる。この免疫反応を強化することによって癌の生育が妨げられる場合、または癌が宿主から排除される場合に、この生体反応を抗腫瘍免疫反応と呼ぶ。
9. 細菌における薬剤耐性は、染色体上の遺伝子に支配されるものもあるが、腸内細菌やブドウ球菌などの臨床分離株、ことに多剤耐性株では、耐性遺伝子がプラスミド上に存在している場合が多い。
10. RNAポリメラーゼが一つのオペロンのプロモーターから転写を開始した時、そのオペロンの末端に存在する転写終結部位が欠失している場合、転写終結が正常に行われず、読み過ごし転写という現象が起こる。

| | | |
|---|---|---|
| 一過性 | イッカセイ | transitory |
| 細動脈 | サイドウミャク | arteriole |
| 収縮 | シュウシュク | contraction |
| ついで | | after that |
| 毛細血管 | モウサイケッカン | capillary |
| 細静脈 | サイジョウミャク | venule |
| 浸出 | シンシュツ | exudation |
| 挿入 | ソウニュウ | insertion |
| 越える | こえる | to transcend |
| 助け | たすけ | aid, assistance |
| 借りる | かりる | to borrow |
| 取り合う | とりあう | to struggle for |
| 器具 | キグ | utensils |
| 腐敗 | フハイ | decay |
| 防ぐ | ふせぐ | to prevent |
| 防腐薬 | ボウフヤク | antiseptic |
| 称する | ショウする | to name, to call |
| 浸潤性 | シンジュンセイ | invasiveness |
| 転移性 | テンイセイ | ability to metastasize |
| 放置する | ホウチする | to leave … as it is, to neglect |
| 妨げる | さまたげる | to hinder |
| 腸内細菌 | チョウナイサイキン | enteric bacteria |
| ブドウ球菌 | ブドウキュウキン | *Staphylococcus* |
| 臨床分離 | リンショウブンリ | clinical isolation |
| ことに | | particularly |

**Ex. 9.5 Additional dictionary entries**

( ) 活性汚泥
( ) 逆位
( ) 逆方向の遺伝学
( ) 酵素の精製
( ) 酵素の分子活性
( ) 半保存的複製
( ) 非相同的組換え
( ) 非反復配列
( ) 表現型混合
( ) 付加突然変異

1. 一連のDNAの配列が染色体の中で向きを反転させた状態で存在すること。
2. 遺伝物質としてのDNA(RNA)に余分な塩基対(塩基(が入り込むことで起こる突然変異。
3. 興味対象となる蛋白質の精製から始めてそのアミノ酸を手掛かりとした遺伝子の同定、変異体の作製という方法で進められる遺伝学。
4. 原核生物のゲノムの大部分を占めている、単相体ゲノム中に１～２回しか出現しない塩基配列。
5. 酵素作用の分子的機能を明らかにするだけではなく、生体反応の経路や制御機構を知るために酵素蛋白質を純粋な形で純物質として取出すこと。
6. 単位時間(通常１分(に酵素１分子が変換する基質分子の数。
7. 都市下水や諸種の産業廃水を連続通気かくはんして、それらの含有有機物に対する再資源化能、酸化能の高い種々の好気性細菌を増殖させて得られる泥状の物質。
8. ２種以上のファージを同一菌に感染させた場合、形態形成途中で両者の混合したファージを生じること。
9. 二本鎖DNAが一本鎖DNAに分離され、それぞれの一本鎖DNAを鋳型として相補鎖が合成されるという二本鎖DNAの複製様式。
10. 一つの染色体の内部の二ヶ所の間で起こる組換え、または二つの染色体の非相同部位の間で起こる組換えのこと。

| | | |
|---|---|---|
| 余分 | ヨブン | excess, surplus |
| 入り込む | はいりこむ | to enter |
| 興味対象 | キョウミタイショウ | subject (object) of interest |
| 手掛かり | てがかり | clue, key, lead |
| 機能 | キノウ | function |
| 都市下水 | トシゲスイ | municipal sewage |
| 諸種 | ショシュ | various types |
| 廃水 | ハイスイ | effluent, discharge |
| 通気 | ツウキ | aeration, ventilation |
| 再資源化能 | サイシゲンカノウ | ability to recover materials for reuse |
| 酸化能 | サンカノウ | oxidation potential |
| 泥状 | デイジョウ | mud-like |

**Ex. 9.6 Additional sentence translations**

1. 宿主域変異とはウイルスあるいはファージの変異であり、野生株のウイルスやファージの宿主域に入っていない細胞や細菌を宿主とすることができる変異である。野生株の宿主域を含んだ形で宿主域が広がることが多い。例えば、λファージにおいては、尾部を形成する蛋白質の遺伝子が変異することにより、野生株のλファージが吸着感染できない大腸菌株にも吸着できるようになり、感染が成立する。

2. 蛋白質の構成ペプチド鎖などにおける一次構造が、関連する試料の間で同一、または高い類似性を示す場合、その部分を相同性領域と呼ぶ。免疫グロブリンのポリペプチド鎖ではH鎖およびL鎖ともに定常部が存在し、クラス、サブクラスが同じであれば、類似性の高い構造をもつ。また動物種を越えて相同性領域の見出される場合もある。

3. 自己を構成する抗原に対しては原則として免疫応答が惹起されない。この自己寛容という現象は胎生期ないし新生児期に自己反応性リンパ球クローンが自己抗原に接触して不活性化ないし消失することによって成立すると考えられている。これには例外がある。レンズ硝子体や中枢神経系のミエリンで、これらの抗原はリンパ球から隔離されていて寛容が成立していない。

4. 遺伝的組換えの機構については、主として二つの考え方がある。一つは選択模写説であり、もう一つは切断再結合説である。選択模写説は、DNA複製が鋳型DNAに沿って任意の部位まで進行した後に、他方の相同染色体に移動して複製が続き、組換えDNAの形成に至るという考えである。切断再結合説は、二つの相同な二本鎖DNAの酵素的切断、およびそれに続く再結合によって組換えDNAが形成されるというものである。現在では後者の方が正しいとされている。

5. 軸索が他の神経細胞または効果細胞とシナプス接合をしており、しばしば神経終末がこん棒状となっている。神経終末は神経興奮の伝達を司り、大きさは1～12 μmと幅がある。シナプス前要素(軸索末端)が受容細胞の膜の境界明瞭な斑点に結合し、化学伝達物質をシナプス間隙(15～50 nm)へ放出する。ただし電気緊張性シナプスの間隙は2 nm以下である。神経終末はシナプス前要素に当たる。

6. 肺炎双球菌は肺炎の病因となる細菌の一種である。グラム陽性の双球菌で、多くは条件的嫌気性である。血液寒天によく発育する。L. Pasteur (1881) によって発見された。Griffith (1928) により形質転換が最初にこの菌によって報告された。その後、Avery (1943) により形質転換因子がDNAであることが明かにされたので有名な細菌である。

7. 大部分の発癌物質は、突然変異原性をもつところから、突然変異が化学発癌の根底にあると考えられている。発癌物質のあるものはそのままでは活性をもたず、細菌内で代謝活性化されてDNAと結合する。DNAの傷害は修復されるが、その時の修復によるエラーも突然変異の原因となる。癌遺伝子に生じた突然変異が発癌の引き金となるものと思われる。

8. 殺虫剤は1945以前は無機物と天然物が使われていたが、1945年以降は多くの化学合成品が開発され、現在はほとんど有機殺虫剤となった。有機殺虫剤は化学構造から有機塩素剤、有機リン剤、カルバメート剤、その他に分類される。有機塩素剤は環境残留性や作物残留性を有し,人畜体内に蓄積して慢性毒性を示すおそれがあるので、1971年以来使用の規制されたものが多い。

9. 腫瘍特異抗原は癌細胞の由来した正常母細胞に存在している組織適合性抗原、臓器特異抗原および組織特異抗原とは別に癌化によって新たに癌細胞のみに発現した抗原である。自己の宿主に対して免疫応答を誘導し、腫瘍の拒絶反応を引き起こし得る抗原と規定される。最近の研究では、他種動物の正常膜抗原や同種の中のアロ組織適合性抗原が癌化に際して顔を出す事実が認められてきており、腫瘍の宿主に対する関係においてこれらの抗原は重要な腫瘍特異抗原と考えられる。

10. 暗号には、始めと終りがあって、これがないと遺伝子としては作用しない。つまり、m-RNAに転写されない。図の人工遺伝子にはこの部分が取り付けられていて、実際に遺伝子としての働きをもったものである。読み始めコドンはATGで、読み終りコドンはTAA、TAG、TGAの三種がある。図でも読み終りの部分の上の鎖にTGAのコドンが見られる。もちろん、これらのコドンを単に付けるだけで遺伝子として働くというような単純なものではない。

| | | |
|---|---|---|
| 広がる | ひろがる | to spread out |
| 惹起する | ジャッキする | to bring about |
| 胎生期 | タイセイキ | gestation period |
| 新生児期 | シンセイジキ | period directly after birth |
| 消失 | ショウシツ | disappearance |
| 硝子体 | ショウシタイ | vitreous body |
| 中枢神経系 | チュウスウシンケイケイ | central nervous system |
| ミエリン | | myelin |
| 隔離 | カクリ | separation, isolation |
| -に沿って | にそって | along ... |
| 任意の | ニンイの | arbitrary |
| 軸索 | ジクサク | axon |
| 効果細胞 | コウカサイボウ | effector cell |
| シナプス接合 | シナプスセツゴウ | synaptic junction |
| こん棒状 | コンボウジョウ | cudgel-shaped |
| 興奮 | コウフン | excitation |
| 司る | つかさどる | to administer |
| 幅 | はば | width |
| シナプス前要素 | シナプスゼンヨウソ | presynaptic element |
| 軸索末端 | ジクサクマッタン | axon terminal |
| 境界明瞭な | キョウカイメイリョウな | clearly bounded, clearly defined |
| 斑点 | ハンテン | stigma |
| 化学伝達物質 | カガクデンタツブッシツ | chemical transmitter |
| シナプス間隙 | シナプスカンゲキ | synaptic cleft |
| 電気緊張性シナプス | デンキキンチョウセイシナプス | electrotonic synapse |
| グラム陽性 | グラムヨウセイ | gram positive |
| 双球菌 | ソウキュウキン | diplococcus |
| 報告 | ホウコク | report |
| 根底 | コンテイ | basis, root |
| 有機塩素剤 | ユウキエンソザイ | chlorinated organic compound |
| 有機リン剤 | ユウキリンザイ | organo-phosphorus compound |
| カルバメート剤 | カルバメートザイ | carbamate compound |
| 残留性 | ザンリュウセイ | property in which residue accumulates |
| 人畜体 | ジンチクタイ | bodies of people and animals |
| 慢性 | マンセイ | chronic |
| 規制 | キセイ | regulation |
| 新たに | あらたに | newly, afresh |
| 顔を出す | かおをだす | to appear |
| 取り付ける | とりつける | to install |
| 単に | タンに | simply, merely |
| 単純な | タンジュンな | uncomplicated |

**Translations for Ex. 9.4**

1. If we look at the changes of an inflammation with the passage of time, a transitory contraction of the arterioles occurs immediately following an injury. After that, an expansion of the arterioles and expansions of the capillaries and venules can be seen. An effusion of [blood] plasma components from within the blood vessels toward the injured tissue occurs.
2. A proto-oncogene on a cell chromosome is inserted into the genome of an RNA tumor virus, and changes on the chromosome so that it possesses the capability to undergo a malignant transformation. However, these genes are known to exist widely, transcending the variety of organism, and play a fundamental role in the conduct of life phenomena.
3. Many carcinogens can act as both an initiator of carcinogenesis and a promoter of carcinogenesis. However, there are some carcinogens whose promoting ability is weak and who can only carry out carcinogenesis with the assistance of an auxiliary carcinogen.
4. There are situations in which the catalytic activity of an enzyme is irreversibly inhibited by a variety of chemical compounds. There are many forms of inhibition of enzyme activity; the form in which inhibitor molecules compete with substrate molecules in a struggle for substrate-binding sites is called competitive inhibition.
5. There are many drugs that cause the death of pathogenic microorganisms. A drug used against the bacteria that attach themselves to utensils and other items is known as a disinfectant; we call a drug that we apply to living tissue a germicide; a drug that prevents putrefaction is called an antiseptic. However, there is no strict distinction [among these].
6. The effect of a recessive lethal gene first appears when the gene becomes a homozygote, but a dominant lethal gene will kill all of the individuals that possess that gene.
7. A benign tumor is a tumor that adopts a morphology close to that of the normal cells from which the tumor cells and their arrangement are derived. It is a tumor without the ability to permeate [the body] or the ability to metastasize. The morphology and arrangment of the cells in a malignant tumor differ in various aspects from those of the original cells. A malignant tumor has both the ability to permeate [the body] and the ability to metastasize; if neglected, it is fatal.
8. Because a tumor-specific antigen is foreign matter to a host, a specific immune reaction against this antigen occurs. In instances where strengthening this immune reaction results in obstruction of cancer growth or exclusion of the cancer from the host, this vital reaction is called an antitumor immune reaction.
9. Drug resistance in some bacteria is governed by genes on chromosomes, but for clinically isolated strains such as enteric bacteria or *Staphylococcus*, particularly for strains with multiple drug resistance, [drug] resistance genes often exist on plasmids.
10. Once RNA polymerase has begun transcription from the promoter on one operon, if the transcription termination portion has been deleted from the terminal of the operon, termination of the transcription does not take place normally. Rather, a phenomenon called read-through transcription occurs.

# LESSON 10　第十課

| | | |
|---|---|---|
| 芽 | ガ | sprout, germ, bud |
| | め | sprout, germ, bud |
| | め(ぐむ) | to sprout, bud |
| 筋 | キン | muscle, sinew |
| | すじ | muscle, sinew |
| 骨 | コツ | bone |
| | ほね | bone |
| 失 | シツ | loss; error |
| | うしな(う) | to lose |
| 除 | ジョ | removal; division {math} |
| | のぞ(く) | to remove, exclude |
| 真 | シン | truth, genuineness |
| | ま- | true, genuine |
| 腺 | セン | gland |
| 胚 | ハイ | embryo |
| 排 | ハイ | exclusion, rejection |
| 葉 | ヨウ | leaf; lobe; plane |
| | は | leaf, foliage |

芽　筋
骨　失
除　真
腺　胚
排　葉

# 芽

| | | |
|---|---|---|
| 栄養芽層 | エイヨウガソウ | trophoblast |
| 芽球 | ガキュウ | blast (cell) |
| 芽球化 ★ | ガキュウカ | blast formation |
| 芽球様細胞 | ガキュウヨウサイボウ | blastoid |
| 芽種 ★ | ガシュ | blastoma |
| 芽体 | ガタイ | blastema |
| 芽胞 | ガホウ | spore |
| 肝芽腫 | カンガシュ | hepatoblastoma |
| 極性海綿芽細胞腫 | キョクセイカイメンガサイボウシュ | spongioblastoma polare |
| 巨赤芽球性貧血 | キョセキガキュウセイヒンケツ | megaloblastic anemia |
| 巨大赤芽球 | キョダイセキガキュウ | megaloblast |
| 筋芽細胞 | キンガサイボウ | myoblast |
| 形質芽球 | ケイシツガキュウ | plasmoblast |
| 血管芽細胞 | ケッカンガサイボウ | angioblast |
| 血球芽細胞 | ケッキュウガサイボウ | hemocytoblast |
| 骨芽細胞 | コツガサイボウ | osteoblast |
| 骨髄芽球 | コツズイガキュウ | myeloblast |
| 小麦胚芽 | こむぎハイガ | wheat germ |
| 小麦胚芽凝集素 | こむぎハイガギョウシュウソ | wheat germ agglutinin |
| 出芽 ★ | シュツガ | budding |
| 上衣海綿芽細胞 | ジョウイカイメンガサイボウ | ependymal spongioblast |
| 上衣芽細胞種 | ジョウイガサイボウシュ | ependymoblastoma |
| 松果体芽細胞腫 | ショウカタイガサイボウシュ | pinealoblastoma |
| 腎芽細胞腫 | ジンガサイボウシュ | nephroblastoma |
| 神経芽細胞 | シンケイガサイボウ | neuroblast |
| 神経芽細胞腫 | シンケイガサイボウシュ | neuroblastoma |
| 推定筋芽細胞 | スイテイキンガサイボウ | presumptive myoblast |
| 成虫芽 | セイチュウガ | imaginal bud |
| 赤芽球 | セキガキュウ | erythroblast |
| 繊維芽細胞 | センイガサイボウ | fibroblast |
| 繊維芽細胞増殖因子 | センイガサイボウゾウショクインシ | fibroblast growth factor |
| 象牙芽細胞 | ゾウゲガサイボウ | odontoblast |
| 頂芽優性 | チョウガユウセイ | apical dominance |
| 鉄芽球性貧血 | テツガキュウセイヒンケツ | sideroblastic anemia |
| 肉芽腫 | ニクガシュ | granuloma |
| 麦芽 | バクガ | malt |
| 白芽細胞 | ハクガサイボウ | leukoblast |
| 麦芽糖 | バクガトウ | malt sugar, maltose |
| 発芽 | ハツガ | germination |
| 芽生え | めばえ | seedling |

# 筋

| | | |
|---|---|---|
| 横紋筋 ★ | オウモンキン | striated muscle |
| 筋芽細胞 | キンガサイボウ | myoblast |
| 筋管 | キンカン | myotube |
| 筋緊張性ジストロフィー | キンキンチョウセイジストロフィー | myotonic dystrophy |
| 筋形質 | キンケイシツ | sarcoplasm |
| 筋形成 | キンケイセイ | myogenesis |
| 筋痙攣 | キンケイレン | muscle cramp |
| 筋原繊維 | キンゲンセンイ | myofibril |
| 筋拘縮 | キンコウシュク | muscle contracture |
| 筋細管系 | キンサイカンケイ | sarcotubular system |
| 筋細胞 | キンサイボウ | muscle cell |
| 筋弛緩物質 | キンシカンブッシツ | muscle relaxant |
| 筋収縮 | キンシュウシュク | muscle contraction |
| 筋上皮細胞 | キンジョウヒサイボウ | myoepithelial cell |
| 筋小胞体 | キンショウホウタイ | sarcoplasmic reticulum |
| 筋節 | キンセツ | myotome, sarcomere |
| 筋繊維 | キンセンイ | muscle fiber |
| 筋[繊維]鞘 | キン[センイ]ショウ | sarcolemma |
| 筋蛋白質 | キンタンパクシツ | muscle protein |
| 筋電図 | キンデンズ | electromyogram |
| 筋肉 ★ | キンニク | muscle |
| 筋板 | キンバン | myotome |
| 筋フィラメント | キンフィラメント | myofilament |
| 筋紡錘 | キンボウスイ | muscle spindle |
| 骨格筋 | コッカクキン | skeletal muscle |
| 重症筋無力症 | ジュウショウキンムリョクショウ | myasthenia gravis |
| 除膜筋繊維 | ジョマクキンセンイ | skinned muscle fiber |
| 心筋 | シンキン | myocardium |
| 神経筋シナプス | シンケイキンシナプス | neuromuscular synapse |
| 神経筋接合部 | シンケイキンセツゴウブ | neuromuscular junction |
| 錐外筋繊維 | スイガイキンセンイ | extrafusal fiber |
| 推定筋芽細胞 | スイテイキンガサイボウ | presumptive myoblast |
| 赤筋 | セキキン | red muscle |
| 速筋 | ソクキン | fast twitch muscle |
| 多発性筋炎 | タハツセイキンエン | polymyositis |
| 遅筋 | チキン | slow (twitch) muscle |
| 中分筋 | チュウブンキン | mesomere |
| 白筋 | ハクキン | white muscle |
| 平滑筋 ★ | ヘイカツキン | smooth muscle |
| 平滑筋弛緩物質 | ヘイカツキンシカンブッシツ | smooth muscle relaxant |

# 骨

| | | |
|---|---|---|
| 後骨髄球 | コウコツズイキュウ | metamyelocyte |
| 骨格 | コッカク | skeleton |
| 骨格筋 ★ | コッカクキン | skeletal muscle |
| 骨格蛋白質 | コッカクタンパクシツ | scaffold protein |
| 骨芽細胞 | コツガサイボウ | osteoblast |
| 骨関節炎 | コツカンセツエン | osteoarthritis |
| 骨吸収 | コツキュウシュウ | bone resorption |
| 骨形成不全症 | コツケイセイフゼンショウ | osteogenesis imperfecta |
| 骨原性細胞 | コツゲンセイサイボウ | osteoprogenitor cell |
| 骨細胞 | コツサイボウ | osteocyte |
| 骨小腔 | コツショウクウ | bone lacuna |
| 骨髄 ★ | コツズイ | bone marrow, myeloid |
| 骨髄芽球 | コツズイガキュウ | myeloblast |
| 骨髄球 | コツズイキュウ | myelocyte |
| 骨髄細胞 | コツズイサイボウ | myeloid cell |
| 骨髄細胞系 | コツズイサイボウケイ | myeloid cell series |
| 骨髄死 | コツズイシ | bone marrow death |
| 骨髄腫 | コツズイシュ | myeloma |
| 骨髄腫グロブリン | コツズイシュグロブリン | myeloma globulin |
| 骨髄腫蛋白質 | コツズイシュタンパクシツ | myeloma protein |
| 骨髄性白血病 | コツズイセイハッケツビョウ | myelocytic leukemia |
| 骨髄由来細胞 | コツズイユライサイボウ | bone-marrow-derived cell |
| 骨粗鬆症 | コツソショウショウ | osteoporosis |
| 骨軟化症 | コツナンカショウ | osteomalacia |
| 骨肉腫 | コツニクシュ | osteosarcoma |
| 骨溶解 | コツヨウカイ | osteolysis |
| 細胞骨格 | サイボウコッカク | cytoskeleton |
| 細胞骨格蛋白質 | サイボウコッカクタンパクシツ | cytoskeletal protein |
| 歯槽骨 | シソウコツ | alveolar bone |
| 背骨 | せぼね | backbone, spinal column |
| 前骨髄球 | ゼンコツズイキュウ | promyelocyte |
| 退行性骨関節症 | タイコウセイコツカンセツショウ | degenerative joint disease |
| 多発性骨髄腫 | タハツセイコツズイシュ | multiple myeloma |
| 軟骨 ★ | ナンコツ | cartilage |
| 軟骨細胞 | ナンコツサイボウ | chondrocyte |
| 軟骨腫 | ナンコツシュ | chondroma |
| 軟骨肉腫 | ナンコツニクシュ | chondrosarcoma |
| 軟骨膜 | ナンコツマク | perichondrium |
| 破骨細胞 | ハコツサイボウ | osteoclast |
| 破骨細胞刺激因子 | ハコツサイボウシゲキインシ | osteoclast-activating factor |
| 類骨 | ルイコツ | osteoid |

# 失

| | | |
|---|---|---|
| 運動失調[症] | ウンドウシッチョウ[ショウ] | ataxia |
| 欠失　★ | ケッシツ | deletion |
| 欠失地図作成　★ | ケッシツチズサクセイ | deletion mapping |
| 欠失変異株 | ケッシツヘンイかぶ | deletion mutant |
| 抗原欠失 | コウゲンケッシツ | antigenic deletion |
| 自己失活 | ジコシッカツ | self-quenching |
| 失活　★ | シッカツ | inactivation |
| 失活剤 | シッカツザイ | quencher |
| 挿入失活 | ソウニュウシッカツ | insertional inactivation |
| 熱失活 | ネツシッカツ | thermal inactivation |
| 付加欠失型突然変異 | フカケッシツがたトツゼンヘンイ | addition-deletion mutation |
| 不可避窒素損失 | フカヒチッソソンシツ | obligatory nitrogen loss |

# 除

| | | |
|---|---|---|
| 汚染除去 | オセンジョキョ | decontamination |
| 解除 | カイジョ | removal, cancellation |
| 胸腺摘除 | キョウセンテキジョ | thymectomy |
| 欠除 | ケツジョ | lusting |
| 色素排除試験 | シキソハイジョシケン | dye-exclusion test |
| 重感染排除 | ジュウカンセンハイジョ | superinfection exclusion |
| 除核 | ジョカク | enucleation |
| 除去　★ | ジョキョ | elimination, extirpation |
| 除去修復　★ | ジョキョシュウフク | excision repair |
| 除草剤 | ジョソウザイ | herbicide |
| 除蛋白 | ジョタンパク | deproteinization |
| 除膜筋繊維 | ジョマクキンセンイ | skinned muscle fiber |
| 侵入排除 | シンニュウハイジョ | entry exclusion |
| 切除 | セツジョ | ablation, extirpation |
| 相互排除 | ソウゴハイジョ | mutual exclusion |
| 嚢摘除 | ノウテキジョ | bursectomy |
| 排除　★ | ハイジョ | exclusion |
| 排除限界 | ハイジョゲンカイ | exclusion limit |
| 表面排除 | ヒョウメンハイジョ | surface exclusion |
| 部分排除 | ブブンハイジョ | partial exclusion |
| 抑制解除 | ヨクセイカイジョ | derepression |

## 真

| | | |
|---|---|---|
| 殺真菌薬 | サツシンキンヤク | fungicide |
| 真核細胞 ★ | シンカクサイボウ | eucaryotic cell |
| 真核生物 | シンカクセイブツ | eucaryote |
| 真菌毒 | シンキンドク | fungal toxin |
| 真菌類 ★ | シンキンルイ | *Eumycetes* |
| 真空 | シンクウ | vacuum |
| 真性グロブリン | シンセイグロブリン | euglobulin |
| 真正細菌 | シンセイサイキン | eubacterium |
| 真性腫瘍 | シンセイシュヨウ | true tumor |
| 真正染色質 ★ | シンセイセンショクシツ | euchromatin |
| 真正蝋 | シンセイロウ | true wax |
| 真皮 | シンピ | corium |

## 腺

| | | |
|---|---|---|
| 外分泌腺 ★ | ガイブンピ[ツ]セン | exocrine gland |
| 下垂体-生殖腺系 | カスイタイ-セイショクセンケイ | pituitary-gonadal axis |
| 下垂体性甲状腺機能亢進症 | カスイタイセイコウジョウセンキノウコウシンショウ | pituitary hyperthyroidism |
| 顎下腺 | ガッカセン | submaxillary salivary gland |
| 汗腺 | カンセン | sudoriferous gland, sweat gland |
| 胸腺 | キョウセン | thymus |
| 胸腺依存性抗原 | キョウセンイゾンセイコウゲン | thymus-dependent antigen |
| 胸腺因子 | キョウセンインシ | thymic factor |
| 胸腺核酸 | キョウセンカクサン | thymus nucleic acid |
| 胸腺細胞 | キョウセンサイボウ | thymocyte |
| 胸腺腫 | キョウセンシュ | thymoma |
| 胸腺髄質 | キョウセンズイシツ | thymic medulla |
| 胸腺摘除 | キョウセンテキジョ | thymectomy |
| 胸腺非依存性抗原 | キョウセンヒイゾンセイコウゲン | thymus-independent antigen |
| 胸腺皮質 | キョウセンヒシツ | thymic cortex |
| 胸腺肥大 | キョウセンヒダイ | thymic hyperplasia |
| 胸腺由来細胞 | キョウセンユライサイボウ | thymus-derived cell |
| 絹糸腺 | ケンシセン | silk gland |
| 甲状腺 ★ | コウジョウセン | thyroid (gland) |
| 甲状腺炎 | コウジョウセンエン | thyroiditis |
| 甲状腺機能低下症 | コウジョウセンキノウテイカショウ | hypothyroidism |
| 甲状腺機能亢進症 | コウジョウセンキノウコウシンショウ | hyperthyroidism |

| | | |
|---|---|---|
| 甲状腺刺激 | コウジョウセンシゲキ | thyroid stimulation |
| 甲状腺刺激抗体 | コウジョウセンシゲキコウタイ | thyroid-stimulating antibody |
| 甲状腺腫 | コウジョウセンシュ | goiter |
| 甲状腺濾胞 | コウジョウセンロホウ | thyroid follicle |
| 自己免疫性甲状腺炎 | ジコメンエキセイコウジョウセンエン | autoimmune thyroiditis |
| 松果腺 | ショウカセン | pineal gland |
| 消化腺 | ショウカセン | digestive gland |
| 生殖腺 | セイショクセン | gonad |
| 性腺刺激物質 | セイセンシゲキブッシツ | gonadotropic substance |
| 性腺刺激ホルモン | セイセンシゲキホルモン | gonadotropic hormone |
| 腺下垂体 | センカスイタイ | adenohypophysis |
| 前胸腺 | ゼンキョウセン | prothoracic gland |
| 前胸腺細胞 | ゼンキョウセンサイボウ | prothymocyte |
| 前胸腺刺激ホルモン | ゼンキョウセンシゲキホルモン | prothoracicotropic hormone |
| 前胸腺ホルモン | ゼンキョウセンホルモン | prothoracic gland hormone |
| 腺細胞 | センサイボウ | glandular cell |
| 腺腫 | センシュ | adenoma |
| 腺性 | センセイ | glandular |
| 腺房 | センボウ | acinus |
| 腺房細胞 | センボウサイボウ | acinar cell |
| 前立腺 | ゼンリツセン | prostate (gland) |
| 唾液腺 | ダエキセン | salivary gland |
| 唾液腺染色体 | ダエキセンセンショクタイ | salivary (gland) chromosome |
| 多腺性内分泌障害 | タセンセイナイブンピ[ツ]ショウガイ | polyendocrinopathy |
| 多発性内分泌腺腫症 | タハツセイナイブンピ[ツ]センシュショウ | multiple endocrine adenomatosis |
| 内分泌腺 ★ | ナイブンピ[ツ]セン | endocrine gland |
| 乳腺 | ニュウセン | mammary gland |
| 乳腺刺激ホルモン | ニュウセンシゲキホルモン | mammotropic hormone |
| 副甲状腺 | フクコウジョウセン | parathyroid |
| 付属腺 | フゾクセン | accessory gland |
| 閉経婦人尿性腺刺激ホルモン | ヘイケイフジンニョウセイセンシゲキホルモン | human menopausal gonadotropin |
| 房状腺 | ボウジョウセン | acinous gland |
| 慢性甲状腺炎 | マンセイコウジョウセンエン | chronic thyroiditis |
| 離出分泌腺 | リシュツブンピ[ツ]セン | apocrine gland |
| 漏出分泌腺 | ロウシュツブンピ[ツ]セン | eccrine gland |

## 胚

| | | |
|---|---|---|
| 外胚葉 | ガイハイヨウ | ectoderm |
| 原外胚葉性 | ゲンガイハイヨウセイ | ectoblast |
| 原腸胚　★ | ゲンチョウハイ | gastrula |
| 小麦胚芽 | こむぎハイガ | wheat germ |
| 小麦胚芽凝集素 | こむぎハイガギョウシュウソ | wheat germ agglutinin |
| 周縁胞胚 | シュウエンホウハイ | periblastula |
| 神経胚　★ | シンケイハイ | neurula |
| 神経胚形成 | シンケイハイケイセイ | neurulation |
| 桑実胚 | ソウジツハイ | morula |
| 中胚葉 | チュウハイヨウ | mesoderm |
| 中胚葉細胞 | チュウハイヨウサイボウ | mesodermal cell |
| 中胚葉母細胞 | チュウハイヨウボサイボウ | mesoblast |
| 内胚乳 | ナイハイニュウ | endosperm |
| 内胚葉 | ナイハイヨウ | endoderm |
| 鶏胚 | にわとりハイ | chick embryo |
| 胚形成 | ハイケイセイ | embryogenesis |
| 胚結節 | ハイケッセツ | embryoblast |
| 胚細胞 | ハイサイボウ | germ cell |
| 胚珠 | ハイシュ | ovule |
| 胚状細胞 | ハイジョウサイボウ | goblet cell |
| 胚中心 | ハイチュウシン | germinal center |
| 胚培養 | ハイバイヨウ | embryo culture |
| 胚盤胞 | ハイバンホウ | blastocyst |
| 胚盤葉 | ハイバンヨウ | blastoderm |
| 胚盤葉下層 | ハイバンヨウカソウ | hypoblast |
| 胚葉 | ハイヨウ | germ layer |
| 盤状胞胚 | バンジョウホウハイ | discoblastula |
| 不定胚 | フテイハイ | adventitious embryo |
| 胞胚　★ | ホウハイ | blastula |
| 胞胚腔 | ホウハイクウ | blastocoel |
| 胞胚形成 | ホウハイケイセイ | blastulation |
| 胞胚葉 | ホウハイヨウ | blastoderm |
| 無腔胞胚 | ムクウホウハイ | stereoblastula |
| 有腔胞胚 | ユウクウホウハイ | coeloblastula |

## 排

| | | |
|---|---|---|
| 色素排除試験 | シキソハイジョシケン | dye-exclusion test |
| 重感染排除 | ジュウカンセンハイジョ | superinfection exclusion |
| 侵入排除 | シンニュウハイジョ | entry exclusion |
| 相互排除 | ソウゴハイジョ | mutual exclusion |
| 尿酸排出 | ニョウサンハイシュツ | uricotelism |

| | | |
|---|---|---|
| 尿素排出 | ニョウソハイシュツ | ureotelism |
| 排出 ★ | ハイシュツ | excretion |
| 排除 ★ | ハイジョ | exclusion |
| 排除限界 | ハイジョゲンカイ | exclusion limit |
| 排卵 ★ | ハイラン | ovulation |
| 排卵誘発剤 | ハイランユウハツザイ | ovulation-inducing agent |
| 排卵抑制剤 | ハイランヨクセイザイ | ovulation inhibitor |
| 表面排除 | ヒョウメンハイジョ | surface exclusion |
| 部分排除 | ブブンハイジョ | partial exclusion |

# 葉

| | | |
|---|---|---|
| 外胚葉 | ガイハイヨウ | ectoderm |
| 下垂体後葉 | カスイタイコウヨウ | posterior pituitary |
| 下垂体後葉細胞 | カスイタイコウヨウサイボウ | pituicyte |
| 下垂体前葉 | カスイタイゼンヨウ | anterior pituitary |
| 肝小葉 | カンショウヨウ | hepatic lobule |
| 原外胚葉性 | ゲンガイハイヨウセイ | ectoblast |
| 後葉 | コウヨウ | infundibular process (referring to pituitary) |
| 抗葉酸剤 | コウヨウサンザイ | antifolate |
| 後葉ホルモン | コウヨウホルモン | posterior lobe hormone |
| 子葉因子 | シヨウインシ | cotyledon factor |
| 神経葉 | シンケイヨウ | neural lobe |
| 中胚葉 | チュウハイヨウ | mesoderm |
| 中胚葉細胞 | チュウハイヨウサイボウ | mesodermal cell |
| 中胚葉母細胞 | チュウハイヨウボサイボウ | mesoblast |
| 中葉 | チュウヨウ | intermediate lobe |
| 中葉ホルモン | チュウヨウホルモン | intermediate lobe hormone |
| 内胚葉 | ナイハイヨウ | endoderm |
| 胞胚葉 | ホウハイヨウ | blastoderm |
| 葉酸 | ヨウサン | folic acid |
| 葉酸塩 | ヨウサンエン | folate |
| 葉鞘 | ヨウショウ | sheath |
| 葉肉細胞 | ヨウニクサイボウ | mesophyll cell |
| 葉脈 | ヨウミャク | vein |
| 葉緑素 | ヨウリョクソ | chlorophyll |
| 葉緑体 | ヨウリョクタイ | chloroplast |
| 葉裂 | ヨウレツ | delamination |
| 胚盤葉 | ハイゲンヨウ | blastoderm |
| 胚盤葉下層 | ハイバンヨウカソウ | hypoblast |
| 胚葉 | ハイヨウ | germ layer |

## EXERCISES

### Ex. 10.1 Matching Japanese and English terms

( ) 汚染除去
( ) 筋芽細胞
( ) 筋繊維
( ) 欠失地図作成
( ) 骨格筋
( ) 骨芽細胞

( ) 骨髄由来細胞
( ) 真核生物
( ) 真菌類
( ) 神経葉
( ) 唾液腺
( ) 内分泌腺

( ) 熱失活
( ) 排除限界
( ) 胚培養
( ) 排卵
( ) 胞胚葉

1. blastoderm
2. bone-marrow-derived cell
3. decontamination
4. deletion mapping
5. embryo culture
6. endocrine gland
7. eucaryote
8. Eumycetes
9. exclusion limit
10. muscle fiber
11. myoblast
12. neural lobe
13. osteoblast
14. ovulation
15. salivary gland
16. skeletal muscle
17. thermal inactivation

### Ex. 10.2 KANJI with similar structural elements

Look carefully at each of the two KANJI on the left, and note which structural element is common to both. Combine each KANJI on the left with the appropriate KANJI on the right to make a meaningful JUKUGO. Each technical term that contains one or more of the 100 KANJI introduced in this book can be found in the vocabulary lists for those KANJI. Other terms can be found in one of the supplementary vocabulary lists, including Lesson 0.

| | (1) | (2) | | |
|---|---|---|---|---|
| 1. | (1) 芽 | (2) 葉 | 出( ) | ( )緑体 |
| 2. | (1) 癌 | (2) 瘍 | 抗( )剤 | 腫( )抗原 |
| 3. | (1) 殺 | (2) 疫 | ( )菌性 | 免( )抑制 |
| 4. | (1) 死 | (2) 化 | ( )学発癌 | 腫瘍壊( )因子 |
| 5. | (1) 失 | (2) 株 | 挿入( )活 | 野生( ) |
| 6. | (1) 失 | (2) 天 | 寒( )内浮遊培養法 | 付加欠( )型突然変異 |
| 7. | (1) 除 | (2) 阻 | 競合( )害剤 | 抑制解( ) |
| 8. | (1) 真 | (2) 直 | ( )正染色質 | ( )接型発癌物質 |
| 9. | (1) 筋 | (2) 節 | 神経( )接合部 | 退行性骨関( )症 |
| 10. | (1) 腺 | (2) 線 | 外分泌( ) | 放射( )殺菌 |
| 11. | (1) 致 | (2) 放 | ( )射線突然変異生成 | 条件( )死変異 |
| 12. | (1) 読 | (2) 続 | 解( ) | 不連( )複製 |
| 13. | (1) 排 | (2) 非 | ( )競合的拮抗薬 | ( )卵抑制剤 |
| 14. | (1) 胚 | (2) 不 | ( )形成 | ( )全症候群 |

**Ex. 10.3 Matching Japanese technical terms with definitions**

Read each definition carefully, and then choose the appropriate technical term. Words that you have not yet encountered are listed following the definitions.

( ) 外分泌腺　( ) 骨細胞　( ) 真核生物　( ) 胚培養　( ) 発芽
( ) 骨格筋　( ) 除去修復　( ) 熱欠失　( ) 排卵　( ) 葉緑体

1. 種子中の胚、胞子、芽などが生育を始めること。
2. 骨格を動かす筋肉で、脊椎動物では骨に付着し、節足動物ではクチクラを動かすもの。
3. 完成された骨の中の主な細胞で、石灰化した細胞間質中の小腔に存在するもの。
4. 蛋白質などの生体高分子の機能が熱を加えることによって失われる現象。
5. 暗修復の一種で、DNAに生じた損傷に対する修復過程のなかでは速やかに行われるもの。
6. 核膜によって仕切られた核構造を細胞内にもつ生物群の総称。
7. 導管を通じ、体表または管腔の内部に分泌物を排出する腺。
8. 動物の体外受精後、受精卵を移植可能な状態にまで生体外または卵管などを利用し、生体内で培養する技術。
9. 一定の成熟段階に達した卵が卵巣から排出されること。
10. 植物細胞中に存在する光合成を行う細胞小器官。

| | | | | | |
|---|---|---|---|---|---|
| 脊椎動物 | セキツイドウブツ | vertebrates | 管腔 | カンクウ | lumen, canal |
| 完成 | カンセイ | completion | 体外受精 | タイガイジュセイ | external fertilization |
| 石灰化 | セッカイカ | calcification | | | |
| 導管 | ドウカン | duct | 卵巣 | ランソウ | ovary |

**Ex. 10.4 Sentence translations**

Read each sentence carefully, and then translate it. Words that you have not yet encountered are listed following the sentences.

1. 出芽は分裂とともに単細胞生物および下等後生動物に多く見られる無性生殖の一種で、個体の体壁の一部に芽体を生じ、これが次第に成長して元の個体と同様な形態となる。
2. 平滑筋の収縮は遅く、一回の収縮に数十秒を要することがあるが、著しく伸長したり短縮することができ、維持や保持の機能を果たす。
3. 骨髄腫は骨髄細胞から発生する腫瘍の総称であるが、その大部分は抗体生成能をもつ形質細胞腫である。骨髄内の各所に多発する傾向があり、このような場合には多発性骨髄腫と呼ばれる。
4. 欠失変異株と各種の突然変異株を用いて相互の染色体上での位置を分子遺伝学的に決定することを欠失地図作成といい、遺伝子微細構造を遺伝学的に解析する場合には信頼性の高い方法となる。
5. 接触型除草剤は除草剤が接触した部分だけを枯らすのに対し、移動型除草剤は植物体内を移動し、作用点に到達後、殺草作用を示めす。殺草効果がでるまでは時間を要するが、殺草効果は大きい。

6. 真核細胞の染色体は紡錘糸の付着点となる動原体をもち、その両側を染色体の腕と呼び、長さに応じて短腕(p)、長腕(q)とする。
7. 内分泌腺の合成・分泌する化学物質は一般にホルモンと呼ばれ、血流によって遠く離れた場所に運ばれて特定の標的器官に作用を及ぼす。
8. 卵割中の受精卵はふつう、胚とよぶことは少ないが、それ以後のものは桑実胚、胞胚、原腸胚、さらに脊椎動物では神経胚とよばれる時期に分けられる。
9. 外胚葉からは表皮・神経系・感覚器官など、中胚葉からは筋肉・腎臓その他多くの器官や組織、内胚葉からは消化管とその付属腺などが生じる。
10. 生体が代謝の不用産物や体内に生じた有害物質を体外または代謝系外に排除する過程を排出という。体内のpH・浸透圧を一定に保ち、内部環境の恒常性を維持するのも排出の意義の一部である。

| | | | | | |
|---|---|---|---|---|---|
| 下等 | カトウ | lower order | 枯らす | からす | to kill (vegetation) |
| 後生動物 | コウセイドウブツ | metazoans | 到達 | トウタツ | arrival, attainment |
| 体壁 | タイヘキ | body wall | 紡錘糸 | ボウスイシ | spindle fiber |
| 同様な | ドウヨウな | same type | 腕 | うで | arm |
| 秒 | ビョウ | second (of time) | 短腕 | タンワン | short arm |
| 短縮 | タンシュク | contraction | 長腕 | チョウワン | long arm |
| 微細構造 | ビサイコウゾウ | fine structure | 腎臓 | ジンゾウ | kidney |
| 信頼性 | シンライセイ | reliability | 浸透圧 | シントウアツ | osmotic pressure |
| 接触型 | セッショクがた | contact-type | 意義 | イギ | meaning, significance |

### Ex. 10.5 Translating titles of journal articles

The titles of journal articles are often written in a condensed style. Read each title carefully, and then translate it. Words that you have not yet encountered are listed following the titles.

1. ゲノム特異的反復配列を用いた植物染色体の研究
2. 共焦点レーザー顕微鏡: 生理学への応用
3. 血小板と粘着性蛋白質
4. 中枢神経系内での興奮性アミノ酸の生合成・代謝
5. 培養肝細胞の構造と機能
6. 細胞膜におけるカルシウム輸送の機構
7. 糖蛋白質糖鎖の生合成と酵素
8. ペルオキシソーム形成機構の研究とペルオキシソーム欠損症病因
9. レトロウイルス感染症の臨床的研究
10. 内分泌細胞の免疫電子顕微鏡法
11. 染色体工学から遺伝子へ: 染色体微細切断法と微量クローニング法
12. 植物の生殖プロトプラストの細胞融合
13. 走査型トンネル顕微鏡による有機薄膜の観察
14. 食細胞の活性酸素生成系と生体防御上の意義
15. 細胞の分裂寿命と不死化

| | | |
|---|---|---|
| 共焦点 | キョウショウテン | confocal |
| 血小板 | ケッショウバン | platelet |
| 粘着性 | ネンチャクセイ | adhesive, sticky |
| 興奮性 | コウフンセイ | excitatory |
| 肝細胞 | カンサイボウ | hepatocyte |
| 薄膜 | ハクマク | thin film |
| 食細胞 | ショクサイボウ | phagocyte |
| 寿命 | ジュミョウ | life (span) |

**Translations for Ex. 10.4**

1. Along with division, budding is a form of asexual propagation that is often observed in unicellular organisms and lower order metazoans. A blastema forms in one portion of the body wall of the individual. This blastema gradually grows and assumes the same kind of morphology as the original individual.
2. The contractions of smooth muscles are slow; in some cases several dozen seconds are required for one contraction. However, smooth muscles can extend and contract to a remarkable degree, and they carry out functions such as support and preservation.
3. Myeloma is the general term for tumors that develop from myeloid cells; the majority are plasmocytoma that carry the ability to produce antibodies. There is a tendency for several tumors to appear at various locations within the bone marrow, and in such situations they are called multiple myeloma.
4. The process of using deletion mutants and various other forms of mutants to determine via molecular genetics positions on common chromosomes is called deletion mapping. It is a highly reliable method for genetically analyzing the fine structure of genes.
5. In contrast to a contact-type herbicide, which only kills the portions [of the vegetation] that it contacts, a mobile herbicide travels within the body of a plant and displays its herbicadal action after arrival at the point of action. Time is necessary for the herbicidal effect to appear, but the herbicidal effect is great.
6. The chromosomes of a eucaryote possess centromeres, which serve as points of attachment for the spindle fibers. Those two ends are called the arms of the chromosomes; depending upon their length we denote the chromosomes as long armed (p) or short armed (q).
7. The chemical substances synthesized and secreted by the endocrine glands are generally called hormones. They are carried by the flow of blood to distant parts of the body, and they exert their effect on specific target organs.
8. A fertilized egg that is undergoing cleavage is seldom called an embryo; the subsequent development is divided into stages called the morula, the blastula, the gastrula, and for vertebrates the neurula.
9. The epidermis, nervous system, sensory organs, and so on, arise from the ectoderm. Muscles, the kidney, and many other organs and types of tissue arise from the mesoderm. The alimentary canal, its associated glands, and so forth, arise from the endoderm.
10. The name excretion is given to the process by which an organism excludes outside the body or outside the metabolic system unnecessary products of metabolism or harmful substances created within the body. Maintaining constant pH and constant osmotic pressure within the body and supporting the homeostasis of the internal environment are each part of the significance of excretion.

## APPENDIX A: SUPPLEMENTARY KANJI LIST

In Chapters 5–20 of BTJ you learned 365 KANJI, and in this volume you have learned an additional 100 KANJI. If you would like to round out your list to 500 KANJI, you could learn the following 35 KANJI, each of which appears ten or more times in this book. For the sake of brevity, you will find here only the most important KUN and ON readings (marked with ▲) and JUKUGO for each KANJI.

| | | | |
|---|---|---|---|
| 悪 | ▲悪 | アク | evil |
| | 悪性腫瘍 | アクセイシュヨウ | malignant tumor |
| | 癌性悪液質 | ガンセイアクエキシツ | cancerous cachexia |
| 医 | ▲医 | イ | medicine, doctor |
| | 医学 | イガク | medicine, medical science |
| | 医師 | イシ | doctor, physician |
| | 医療 | イリョウ | medical treatment |
| 意 | ▲意 | イ | mind, heart; intention; thought |
| | 意義 | イギ | meaning, significance |
| | 意味 | イミ | meaning |
| | 任意の | ニンイの | arbitrary, random |
| 維 | ▲維 | イ | tieing; rope |
| | 維持 | イジ | support, maintain |
| | 筋繊維 | キンセンイ | muscle fiber |
| | 繊維芽細胞 | センイガサイボウ | fibroblast |
| 完 | ▲完 | カン | completion |
| | 完結 | カンケツ | completion, conclusion |
| | 完成 | カンセイ | completion |
| | 完全 | カンゼン | completeness |
| 肝 | ▲肝 | カン | liver |
| | 肝炎 | カンエン | hepatitis |
| | 肝癌 | カンガン | hepatic carcinoma, hepatoma |
| | 肝臓 | カンゾウ | liver |
| 義 | ▲義 | ギ | meaning; integrity; justice |
| | 意義 | イギ | meaning, significance |
| | 狭義 | キョウギ | narrow sense, narrow meaning |
| | 広義 | コウギ | broad sense, broad meaning |
| | 定義 | テイギ | definition |

| | | | |
|---|---|---|---|
| 激 | ▲激 | ゲキ | excitement, agitation |
| | ▲激しい | はげ(しい) | violent |
| | 急激に | キュウゲキに | abruptly, precipitously |
| | 刺激 | シゲキ | stimulus |
| | 泌乳刺激ホルモン | ヒ[ツ]ニュウシゲキホルモン | lactogenic hormone |
| 候 | ▲候 | コウ | season, weather |
| | 後天性免疫不全症候群 | コウテンセイメンエキフゼンショウコウグン | acquired immune deficiency syndrome (AIDS) |
| | 症候群 | ショウコウグン | syndrome |
| 持 | ▲持 | ジ | holding, maintaining |
| | ▲持つ | も(つ) | to have, to possess |
| | 維持 | イジ | support, maintain |
| | 支持 | シジ | support |
| | 保持 | ホジ | preservation |
| 弱 | ▲弱 | ジャク | weakness |
| | ▲弱い | よわ(い) | weak |
| | 弱酸 | ジャクサン | weak acid |
| | 弱毒化 | ジャクドクカ | attenuation |
| | 弱有害遺伝子 | ジャクユウガイイデンシ | mildly deleterious gene |
| 手 | ▲手 | シュ | hand; means |
| | ▲手 | て | hand; means |
| | 手段 | シュダン | means, measure |
| | 手法 | シュホウ | technique |
| | 手掛かり | てがかり | clue, key, lead |
| 取 | ▲取 | シュ | taking, obtaining |
| | ▲取る | と(る) | to take, to obtain |
| | 取り扱う | とりあつかう | to treat, to deal with |
| | 取り出す | とりだす | to take (out) |
| | 取り込む | とりこむ | to incorporate |
| 純 | ▲純 | ジュン | purity |
| | 純系 | ジュンケイ | pure line |
| | 純培養 | ジュンバイヨウ | pure culture |
| | 単純な | タンジュンな | simple, uncomplicated |

| | | | |
|---|---|---|---|
| 初 | ▲初 | ショ | beginning; first |
| | ▲初めて | はじ(めて) | for the first time |
| | 最初 | サイショ | initial |
| | 初期 | ショキ | early stage |
| | 初代 | ショダイ | first generation |
| 腎 | ▲腎 | ジン | kidney |
| | 腎炎 | ジンエン | nephritis |
| | 腎臓 | ジンゾウ | kidney |
| | 腎皮質 | ジンヒシツ | renal cortex |
| 髄 | ▲髄 | ズイ | marrow, pith |
| | 骨髄 | コツズイ | bone marrow |
| | 骨髄腫 | コツズイシュ | myeloma |
| 絶 | ▲絶 | ゼツ | extremity |
| | ▲絶えず | た(えず) | ceaselessly |
| | 移植片拒絶反応 | イショクヘンキョゼツハンノウ | graft rejection |
| | 絶対嫌気性細菌 | ゼッタイケンキセイサイキン | strictly anaerobic bacterium |
| | 絶対好気性細菌 | ゼッタイコウキセイサイキン | strictly aerobic bacterium |
| 繊 | ▲繊 | セン | slender, fine |
| | 筋繊維 | キンセンイ | muscle fiber |
| | 繊維芽細胞 | センイガサイボウ | fibroblast |
| | 微小繊維 | ビショウセンイ | microfilament |
| 臓 | ▲臓 | ゾウ | viscera, bowels |
| | 肝臓 | カンゾウ | liver |
| | 腎臓 | ジンゾウ | kidney |
| | 臓器 | ゾウキ | internal organ |
| 損 | ▲損 | ソン | loss; disadvantage |
| | ▲損う | そこな(う) | to harm; to lose |
| | 欠損 | ケッソン | defect |
| | 欠損症 | ケッソンショウ | deficiency |
| | 損傷 | ソンショウ | damage, injury |
| 短 | ▲短 | タン | shortness, brevity |
| | ▲短い | みじか(い) | short |
| | 短期培養 | タンキバイヨウ | short-term culture |
| | 短縮 | タンシュク | contraction |

| | | | |
|---|---|---|---|
| 団 | ▲団 | ダン | group, corps |
| | 細胞集団 | サイボウシュウダン | cell population |
| | 集団遺伝学 | シュウダンイデンガク | population genetics |
| | 集団培養 | シュウダンバイヨウ | mass culture |
| 虫 | ▲虫 | チュウ | insect, worm |
| | ▲虫 | むし | insect, worm |
| | 寄生虫 | キセイチュウ | parasite |
| | 殺虫剤 | サッチュウザイ | insecticide |
| | 線虫 | センチュウ | nematode |
| 腸 | ▲腸 | チョウ | intestines, bowels |
| | 大腸菌 | ダイチョウキン | *Escherichia coli* (*E. coli*) |
| | 腸管毒 | チョウカンドク | enterotoxin |
| | 腸内細菌 | チョウナイサイキン | enteric bacteria |
| 肉 | ▲肉 | ニク | flesh, muscle; meat |
| | 筋肉 | キンニク | muscle |
| | 骨肉腫 | コツニクシュ | osteosarcoma |
| | 肉腫 | ニクシュ | sarcoma |
| 乳 | ▲乳 | ニュウ | milk; the breasts |
| | 乳腺 | ニュウセン | mammary gland |
| | 泌乳 | ヒ[ツ]ニュウ | lactation |
| | 哺乳動物 | ホニュウドウブツ | mammals |
| 脳 | ▲脳 | ノウ | brain |
| | 大脳新皮質 | ダイノウシンヒシツ | cerebral neocortex |
| | 大脳皮質 | ダイノウヒシツ | cerebral cortex |
| | 脳炎 | ノウエン | encephalitis |
| 肺 | ▲肺 | ハイ | lung |
| | 肺炎 | ハイエン | pneumonia |
| | 肺炎双球菌 | ハイエンソウキュウキン | *Diplococcus pneumoniae* |
| | 肺癌 | ハイガン | lung cancer |
| | 肺臓 | ハイゾウ | lung |

| | | | |
|---|---|---|---|
| 浮 | ▲浮 | フ | floating |
| | ▲浮かぶ | う(かぶ) | to float |
| | 寒天内浮遊培養法 | カンテンナイフユウバイヨウホウ | agar suspension culture |
| | 浮腫 | フシュ | edema |
| | 浮遊[細胞]培養 | フユウ[サイボウ]バイヨウ | suspension (cell) culture |
| 保 | ▲保 | ホ | keeping, maintaining |
| | ▲保つ | たも(つ) | to preserve, to maintain |
| | 保護 | ホゴ | protection |
| | 保持 | ホジ | preservation |
| | 保存 | ホゾン | stock; conservation |
| 命 | ▲命 | ミョウ・メイ | life; command |
| | ▲命 | いのち | life |
| | 寿命 | ジュミョウ | life(span) |
| | 生命 | セイメイ | life |
| 滅 | ▲滅 | メツ | ruination; extermination |
| | 死滅 | シメツ | death, extinction |
| | 消滅 | ショウメツ | disappearance |
| | 滅菌 | メッキン | sterilization |
| 良 | ▲良 | リョウ | good |
| | ▲良い | よ(い) | good |
| | 改良 | カイリョウ | improvement |
| | 優良 | ユウリョウ | superior |
| | 良性腫瘍 | リョウセイシュヨウ | benign tumor |
| 療 | ▲療 | リョウ | healing, cure |
| | 遺伝子療法 | イデンシリョウホウ | gene therapy |
| | 医療 | イリョウ | medical treatment |
| | 化学療法剤 | カガクリョウホウザイ | chemotherapeutic agent |

## APPENDIX B: ON-KUN INDEX FOR THE KANJI EMPHASIZED IN THIS BOOK

The number after a KANJI gives the lesson in which that KANJI is introduced; "S" stands for the supplementary list in Appendix A. ON readings are given on this page in the GOJUU-ON order; KUN readings are listed on the following page.

アク　悪[S]
イ　医[S] 意[S] 維[S]
イキ　域[8]
イン　因[1]
エイ　栄[2] 泳[2]
エキ　疫[1]
エン　遠[1] 炎[9]
ガ　画[7] 芽[10]
ガイ　害[2]
カク　画[7] 拡[7] 獲[8]
カン　還[1] 寒[3] 感[5] 緩[6] 寛[8] 完[S] 肝[S]
ガン　癌[9]
ギ　義[S]
キョウ　鏡[4] 競[9]
ギョウ　凝[5]
キン　筋[10]
グン　群[7]
ケイ　蛍[4]
ゲキ　激[S]
ケツ　欠[7]
ケン　嫌[3] 顕[4]
コ　呼[3] 己[8]
コウ　降[1] 好[3] 候[S]
コツ　骨[10]
サ　鎖[6]
ザ　座[7]
サツ　殺[9]
ザツ　雑[4]
シ　止[3] 紫[4] 枝[5] 指[6] 始[7] 死[9]

ジ　持[S]
シキ　識[6]
シツ　失[10]
シャ　謝[3] 写[7]
ジャク　弱[S]
シュ　主[8] 腫[9] 手[S] 取[S]
シュウ　修[4] 終[6]
シュク　宿[8]
ジュン　純[S]
ショ　初[S]
ジョ　除[10]
ショウ　照[2] 障[2] 症[7]
ショク　蝕[4]
シン　親[6] 真[10]
ジン　腎[S]
ズイ　髄[S]
セイ　清[2]
セツ　節[3] 切[8]
ゼツ　絶[S]
セン　旋[5] 腺[10] 繊[S]
ソ　阻[2]
ソウ　走[6]
ゾウ　臓[S]
ソク　促[5]
ソン　損[S]
タイ　耐[9]
ダク　濁[3]
ダツ　脱[1]
タン　蛋[1] 短[S]
ダン　団[S]
チ　致[9]

チュウ　虫[S]
チョウ　腸[S]
チン　沈[1]
テン　天[3]
トウ　答[7]
ドク　毒[5] 読[9]
ニク　肉[S]
ニュウ　乳[S]
ノウ　脳[S]
ハイ　胚[10] 排[10] 肺[S]
バイ　培[2]
ヒ　皮[5] 泌[6]
ヒツ　泌[6]
ビョウ　病[5]
ヒン　頻[7]
フ　浮[S]
フク　復[4]
ヘン　片[4]
ホ　補[1] 保[S]
ボ　母[5]
マツ　末[8]
ミョウ　命[S]
メイ　命[S]
メツ　滅[S]
メン　免[1]
ユ　輸[8]
ユウ　融[2] 優[4] 遊[6]
ヨウ　養[2] 瘍[9] 葉[10]
ヨク　抑[6]
リョウ　領[8] 良[S] 療[S]
レイ　冷[3]

| | |
|---|---|
| いたす | 致す[9] |
| いのち | 命[S] |
| いやがる | 嫌がる[3] |
| うかぶ | 浮かぶ[S] |
| うしなう | 失う[10] |
| うつす | 写す[7] |
| うつる | 写る[7] |
| えだ | 枝[5] |
| おえる | 終える[6] |
| おぎなう | 補う |
| おさえる | 抑える[6] |
| おもな | 主な[8] |
| おや | 親[6] |
| おわり | 終り[6] |
| おわる | 終る[6] |
| かがみ | 鏡[4] |
| かく | 欠く[7] |
| かける | 欠ける[7] |
| かた | 片[4] |
| かわ | 皮[5] |
| きらい | 嫌い[3] |
| きらう | 嫌う[3] |
| きる | 切る[8] |
| きれる | 切れる[8] |
| くさり | 鎖[6] |
| こたえ | 答え[7] |
| こたえる | 答える[7] |
| このむ | 好む[3] |
| ころす | 殺す[9] |
| さす | 指す[6] |
| さむい | 寒い[3] |
| しずむ | 沈む[1] |
| しぬ | 死ぬ[9] |
| すえ | 末[8] |
| すき | 好き[3] |
| すく | 好く[3] |
| すじ | 筋[10] |
| そこなう | 損う[S] |
| たえず | 絶えず[S] |
| たえる | 耐える[9] |
| たもつ | 保つ[S] |
| つめたい | 冷たい[3] |
| て | 手[S] |
| とおい | 遠い[1] |
| とおくの | 遠くの[1] |
| とまる | 止まる[3] |
| とめる | 止める[3] |
| とる | 取る[S] |
| にごす | 濁す[3] |
| にごる | 濁る[3] |
| のぞく | 除く[10] |
| は | 葉[10] |
| はげしい | 激しい |
| はしる | 走る[6] |
| はじまり | 始まり[7] |
| はじまる | 始まる[7] |
| はじめて | 初めて[S] |
| はじめる | 始める[7] |
| はは | 母[5] |
| はらす | 腫らす[9] |
| はれる | 腫れる[9] |
| ひえる | 冷える[3] |
| ひやす | 冷やす[3] |
| ふし | 節[3] |
| ほね | 骨[10] |
| ほのお | 炎[9] |
| ま | 真[10] |
| みじかい | 短い[S] |
| むし | 虫[S] |
| むらさき | 紫[4] |
| むれ | 群[7] |
| め | 芽[10] |
| めぐむ | 芽ぐむ[10] |
| もつ | 持つ[S] |
| やしなう | 養う[2] |
| ゆび | 指[6] |
| よい | 良い[S] |
| よぶ | 呼ぶ[3] |
| よむ | 読む[9] |
| よめる | 読める[9] |
| よわい | 弱い[S] |